Human Epigenomics

Carsten Carlberg · Ferdinand Molnár

Human Epigenomics

 Springer

Carsten Carlberg
Institute of Biomedicine
University of Eastern Finland
Kuopio
Finland

Ferdinand Molnár
Institute of Biomedicine
University of Eastern Finland
Kuopio
Finland

ISBN 978-981-13-5660-5   ISBN 978-981-10-7614-5 (eBook)
https://doi.org/10.1007/978-981-10-7614-5

Printed on acid-free paper

This Springer imprint is published by Springer Nature
The registered company is Springer Nature Singapore Pte Ltd.
The registered company address is: 152 Beach Road, #21-01/04 Gateway East, Singapore 189721, Singapore

Preface

The term "epigenetics" was first introduced 75 years ago by Conrad Waddington and is used since then, in order to describe regulatory and information storing mechanisms of specific genes that do not involve any change of their DNA sequence. Epigenetics is closely related to the extensively folded state, in which the genome is packaged, known as chromatin. New genomic tools, which have been developed since the sequencing of the human genome, nowadays allow the genome-wide assessment of, for example, chromatin states and DNA modifications, and led to the discovery of unexpected new epigenetic principles, such as epigenomic memory. This was the start of the field of "epigenomics," the relation of which to human health and disease is discussed in this textbook.

This book aims to summarize, in a condensed form, the role of epigenomics in defining chromatin states that are representative of active genes (euchromatin) and repressed genes (heterochromatin). Moreover, this book discusses the principles of gene regulation, chromatin stability, genomic imprinting and the reversibility of DNA methylation and histone modifications. This information should enable a better understanding of cell type identities and will provide new directions for studies of, for example, (i) cellular reprograming, (ii) the response of chromatin to environmental signals and (iii) epigenetic therapies that can improve or restore human health.

The shift from epigenetics to epigenomics, i.e. from a "single gene" to a "genome-wide" view, significantly enhances the understanding of chromatin accessibility, 3-dimensional (3D) organization of the genome and its consequences on gene expression affecting the phenotype and function of cells. This book will discuss the central importance of epigenomics during embryogenesis and cellular differentiation as well as in the process of aging and the risk for the development of cancer. Moreover, the role of the epigenome as a molecular storage of cellular events not only in the brain but also in the immune system and in metabolic organs will be described.

Over the last decade the terms "epigenetics"/"epigenomics" raised interest not only in biomedicine, such as oncology or nutritional sciences, but also in fields that do not routinely address genetics, such as ecology, physiology and psychology. Some scientists primarily use the term epigenomics to explain changes in gene expression, while others employ the term, in order to refer to transgenerational effects and/or inherited gene expression states. This book aims on to clarify the principles of epigenomics and may be used as a basis for the different disciplines implementing it. For this reason not only biochemists should be aware of the concepts of epigenomics, but all students of biology and medicine would benefit from being familiar with this topic. Thus, a solid understanding of the epigenome should be a fundamental goal of modern life science research and teaching.

The content of the book is based on the lecture course "Molecular Medicine and Genetics" that is given by one of us (C. Carlberg) in different forms since 2002 at the University of Eastern Finland in Kuopio. Thematically, this book is located between our textbooks "Mechanisms of Gene Regulation" (ISBN 978-94-017-7741-4) and "Nutrigenomics" (ISBN 978-3-319-30415-1), studying of which may also be interesting to our readers. The book is sub-divided into three sections and 13 chapters. Following the Introduction (section A), section B will explain the molecular basis of epigenomics, while section C will provide examples for the impact of epigenomics in human health and disease. The lecture course is primarily designed for Master level students of biomedicine, but is also frequented by PhD students as well as by students of other bioscience disciplines. The course and hence this textbook has four major learning objectives. Students should:

1. have detailed understanding of the molecular elements of epigenomics, such as chromatin organization, chemical modifications found in DNA and histones and the nuclear proteins mediating and coordinating these reactions,
2. recognize the key components, mechanisms and processes in epigenomics and the multiple layers of its regulatory complexity in processes, such as cellular differentiation, aging and cancer, and its function in the brain, immune system and metabolic organs,
3. show the ability to analyze the impact of DNA methylation, histone modifications and 3D chromatin organization for human health and disease, and
4. apply knowledge in epigenomics in designing, performing and analyzing respective experiments, such as ChIP-seq, ATAC-seq and single-cell RNA-seq.

We hope the readers will enjoy this rather visual book and get as enthusiastic about epigenomics as the authors are.

October 2017   Carsten Carlberg
Kuopio   Ferdinand Molnár

Acknowledgements

The authors would like to thank Reinhard Bornemann, MD, DrPH, PhD, for extensive proofreading and constructive criticism.

List of Abbreviations

3C	chromosome conformation capture
3D	3-dimensional
5caC	5-carboxylcytosine
5fC	5-formylcytosine
5hmC	5-hydroxymethylcytosine
5mC	5-methylcytosine
α-KG	α-ketoglutarate
βOHB	D-β-hydroxybutyrate
AC	adenylyl cyclase
ACLY	ATP citrate lyase
ACSS2	acyl-CoA synthetase short chain family member 2
AICDA	activation-induced cytidine deaminase
Air	antisense insulin-like growth factor 2 receptor RNA
AML	acute myeloid leukemia
AMPK	AMP-activated protein kinase
ARID	AT-rich interaction domain
ASIP	Agouti signaling protein
ATRX	α-thalassemia/mental retardation syndrome X-linked
BAF	BRG1-associated factor
BDNF	brain-derived neurotrophic factor
BER	base excision repair
BET	bromodomain and extra-terminal
BMI	body mass index
BMI1	BMI1 proto-oncogene, Polycomb ring finger
BMP	bone morphogenetic protein
bp	base pair
CAGE	cap analysis of gene expression
CaMKII	calcium/calmodulin-dependent kinase II
CBFB	core-binding factor subunit-β
CDKN	cyclin-dependent kinase inhibitor
CEBP	CCAAT/enhancer binding protein

CFP1	CXXC finger protein 1
ChIP	chromatin immunoprecipitation
CIMP	CpG island methylator phenotype
CK1	casein kinase 1
CLL	chronic lymphoid leukemia
CLP	common lymphoid progenitor
CML	chronic myeloid leukemia
CMP	common myeloid progenitor
CREBBP	CREB binding protein, also called KAT3A
CTCF	CCCTC binding factor
DBD	DNA-binding domain
DMR	differentially methylated region
DNMT	DNA methyltransferase
DOHaD	developmental origins of health and disease
EMT	epithelial-to-mesenchymal transition
ENCODE	encyclopedia of DNA elements
EP300	E1A binding protein p300, also called KAT3B
EPHA4	EPH Receptor A4
ERK	extracellular regulated kinase
eRNA	enhancer RNA
ES	embryonic stem
EWAS	epigenome-wide association study
EZH	enhancer of zeste homolog
FAD	flavin adenine dinucleotide
FAIRE	formaldehyde-assisted isolation of regulatory elements
FANTOM	Functional ANnotaTion Of the Mammalian genome
FDA	Food and Drug Administration
FISH	fluorescence in situ hybridization
FMR1	fragile X mental retardation 1
FOX	forkhead box
FXN	frataxin
GATA	GATA binding protein
GMP	granulocyte-monocyte progenitor
GPCR	G protein-coupled receptor
gRNA	guide RNA
GTEx	Genotype Tissue Expression
GWAS	genome-wide association study
HAT	histone acetyltransferase
HDAC	histone deacetylase
HGPS	Hutchinson-Gilford progeria syndrome
HMG	high-mobility group protein
HNRNPU	heterogeneous nuclear ribonucleoprotein U
HOTAIR	HOX transcript antisense RNA
HOTTIP	HOXA transcript at the distal tip
HP1	heterochromatin protein 1

HSC	hematopoietic stem cell
HTT	Huntingtin
IAP	intracisternal A particle
ICM	inner cell mass
ICR	imprint control region
IDH	isocitrate dehydrogenase
IGF2	insulin-like growth factor 2
IHEC	International Human Epigenome Consortium
IHH	Indian hedgehog
IL	interleukin
INFG	interferon γ
INO80	INO80 complex subunit
iPS	induced pluripotent stem
ISWI	imitation SWI
IVF	*in vitro* fertilization
KAT	lysine acetyltransferase
kb	kilo base pairs (1,000 bp)
KCNQ1	potassium voltage-gated channel subfamily Q member 1
KDAC	lysine deacetylase
KDM	lysine demethylase
KLF4	Krüppel-like factor 4
KMT	lysine methyltransferase
LAD	lamin-associated domain
LCR	locus control region
LDH	lactate dehydrogenase
LINE	long interspersed element
LOCK	large organized chromatin K9-modification
LPS	lipopolysaccharide
LSD1	lysine specific demethylase 1, also called KDM1A
LTP	long-term potentiation
MAGEA1	melanoma-associated antigen 1
MALAT1	metastasis associated lung adenocarcinoma transcript 1
MAPK	mitogen-activated protein kinases
MAT2A	methionine adenosyltransferase 2A
MBD	methyl-DNA binding domain
mCH	non-CpG methylation
MCPH1	microcephalin
MECOM	MDS1 and EVI1 complex locus
MeCP2	methyl-CpG binding protein 2
MEK	mitogen-activated protein kinase kinase
MEP	megakaryocyte-erythrocyte progenitor
MHC	major histocompatibility complex
MICA	MHC class I polypeptide-related sequence A
miRNA	micro RNA
MLL	mixed lineage leukemia

MPP	multipotent progenitor
mRNA	messenger RNA
MS	multiple sclerosis
MSC	mesenchymal stem cell
MTHFR	methylenetetrahydrofolate reductase
MYF5	myogenic factor 5
MYOD1	myogenic differentiation 1
NAD	nicotinamide adenine dinucleotide
NCOA6	nuclear receptor co-activator 6
NCOR	nuclear receptor co-repressor
ncRNA	non-coding RNA
NK	natural killer
NO	nitric oxide
nt	nucleotides
NuRD	nucleosome remodeling and deacetylase
NuRF	nucleosome remodeling factor
PAMP	pathogen-associated molecular pattern
PAX	paired box
PBMC	peripheral blood mononuclear cell
PcG	Polycomb group
PDC	pyruvate dehydrogenase complex
PDGFRA	platelet-derived growth factor receptor α
PGC	primordial germ cell
PKA	protein kinase A
Pol II	RNA polymerase II
PPAR	peroxisome proliferator-activated receptor
PPARGC1A	proliferator-activated receptor gamma, co-activator 1 alpha
PRC	Polycomb repressive complex
PRDM	PR/SET domain
PRMT	protein arginine methyltransferases
PTSD	post-traumatic stress disorder
QTL	quantitative trait locus
RCOR1	REST co-repressor
RE1	restrictive element 1
REST	RE1-silencing transcription factor
RUNX1	runt-related transcription factor 1
SAH	S-adenosylhomocysteine
SAM	S-adenosylmethionine
SCNT	somatic cell nuclear transfer
SHARP	SMRT/HDAC1-associated repressor protein
SIN3A	SIN3 transcription regulator family member A
SIRT	sirtuins
SLE	systemic lupus erythematosus
SMARCA	SWI/SNF-related matrix-associated actin-dependent regulators of chromatin subfamily A

SNP	single nucleotide polymorphism
SOX2	SRY-box 2
SUV39H	suppressor of variegation 3-9 homolog
SWI/SNF	switching/sucrose non-fermenting
T2D	type 2 diabetes
TAD	topologically associated domain
TAF	TBP-associated factor
TDG	thymine-DNA glycosylase
TET	ten-eleven translocation
T_H	T helper
THA11	THAP domain-containing 11
TRIM	tripartite-motif-containing protein
tRNA	transfer RNA
TrxG	Trithorax group
TSS	transcription start site
UHRF1	ubiquitin-like plant homeodomain and RING finger domain 1
UTR	untranslated region
WNT	Wnt family member
XCI	X chromosome inactivation
Xist	X-inactive specific transcript
ZFP	zinc-finger protein
ZGA	zygotic genome activation

Contents

About the Authors

Prof. Carsten Carlberg Professor for biochemistry at the School of Medicine, Institute of Biomedicine at the University of Eastern Finland. The main research interests of Prof. Carlberg are (epi)genomics of nuclear receptors and their ligands with special focus on vitamin D. So far he published more than 200 papers (H-index 54). Since 2001 he is lecturing yearly courses on epigenetics and gene regulation, which are the basis of this textbook. Together with Dr. Molnár he also published the Springer textbooks "Mechanisms of Gene Regulation" and "Nutrigenomics."

Dr. Ferdinand Molnár Project researcher at the School of Medicine, Institute of Biomedicine at the University of Eastern Finland. The main research interests of Dr. Molnár are the molecular structure of nuclear receptor proteins and their natural and synthetic ligands, on which he published more than 27 papers (H-index 17). Together with Prof. Carlberg he published the Springer textbooks "Mechanisms of Gene Regulation" and "Nutrigenomics."

Part A
Introduction

Chapter 1
What Is Epigenomics?

Abstract Chromatin is a complex of nuclear proteins with genomic DNA that not only allows densely packaging of the genome into the cell nucleus but also tightly controls the access to genes and their regulatory elements, such as promoters and enhancers. The latter process is regulated by methylation of genomic DNA at cytosines and post-translational modifications of histone proteins. Furthermore, gene expression is controlled by DNA looping and other 3D chromatin structures. Epigenetics is defined as the study of changes in gene function that are heritable but do not involve changes in the genome. The genome-wide analysis of epigenetics is referred to as epigenomics. The genome in each cell of a multi-cellular organism is identical and stays reasonably static, while the epigenome is very dynamic, varies from one cell type to the other and can respond to various signaling pathways. Accordingly, also the transcriptome and the proteome of a cell are dynamic and cell-specific.

In this chapter, we will define the term "epigenetics" and its extension "epigenomics" that is based on the application of genome-wide methods. We will provide a first view on the structure of chromatin and its tight link to gene expression by introducing the response of the epigenome to intra- and extra-cellular signals on the level of DNA methylation and histone modifications. Furthermore, we will present chromatin modifying and remodeling enzymes and their role in controlling the different levels of epigenomic/chromatin activity that affect gene expression. In this context, we will introduce the concept of epigenetic landscapes for the visualization of epigenomic changes during cellular differentiation as well as in tumorigenesis. Finally, we will provide an outline of the impact of the epigenome in human health and disease.

Keywords Chromatin · genome · gene expression · central dogma of molecular biology · transcriptome · epigenome · epigenetics · epigenetic landscape · chromatin modifiers · promoter · enhancer · chromatin organization

© Springer Nature Singapore Pte Ltd. 2018

3

C. Carlberg, F. Molnár, *Human Epigenomics*, https://doi.org/10.1007/978-981-10-7614-5_1

1.1   Chromatin and Gene Expression

The vast majority (> 99%) of the human body is composed of terminally differentiated cells that contain a diploid genome normally formed by 2×22 autosomal chromosomes and two sex chromosomes (XX for females or XY for males). Chromosomes are formed of chromatin, which is the macromolecular complex of genomic DNA and nuclear proteins that condenses long DNA molecules (16–85 mm) into the nucleus of a cell (diameter 6–10 µm).

Chromatin is organized into lower-order structures, such as the 10 nm fiber (also referred to as "beads on a string") and higher-order structures, like the 30 nm fiber and the 700 nm mitotic chromosomes. The default state of chromatin, also referred to as heterochromatin, is not accessible to DNA binding transcription factors (Fig. 1.1, *top*). The most densely packing of chromatin is found during the metaphase of mitosis, shortly before the chromosomes are distributed to the daughter cells. This phase has to be short, since at such dense chromatin packing there is no gene transcription, i.e. no flexibility to response to environmental signals.

Thus, chromatin acts as a filter for the access of DNA binding proteins to functional elements on the genome, such as transcription start site (TSS) regions, also

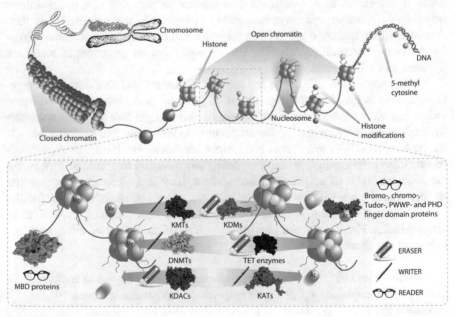

Fig. 1.1 Central role of chromatin. Per nucleosome 147 bp of genomic DNA is wrapped around eight histone proteins, which compacts the genome into chromatin and chromosomes (*top*, Chap. 3). Covalent modifications of histones and genomic DNA, such as methylations, control the accessibility of chromatin to transcription factors and other regulatory proteins. These chromatin marks are introduced by "writers," interpreted by "readers" and can be removed by "erasers" (*bottom*, Chap. 6). The interplay between these nuclear proteins is essential for controlling gene expression

referred to as core promoters, and enhancers (Sect. 1.4). Genes are only expressed when their TSSs are accessible to the basal transcriptional machinery including RNA polymerase II (Pol II) and they are transcribed in an appropriate amount only when their specific enhancer regions are not covered by chromatin and can be recognized by transcription factors.

As indicated by the central dogma of molecular biology (Box 1.1) the initial step in gene expression is the transcription of the genomic DNA of the gene body into RNA, which after splicing and transport from the nucleus to the cytoplasm is translated into protein (Fig. 1.2). Proteins are the "workers" within a cell and basically mediate all functions therein, such as signal transduction, catalysis and control of metabolic reactions, molecule transport and many more. In addition, proteins contribute to the structure and stability of cells and intra-cellular matrices. Therefore, gene expression determines the phenotype, function and developmental state of cell types and tissues, respectively. Gene expression patterns are cell-specific, but can also drastically change after exposure to intra- and extra-cellular signals and in response to pathological conditions, such as infection or cancer.

Box 1.1 The Central Dogma of Molecular Biology

The dogma indicates a clear direction in the flow of information from DNA to RNA to protein (Fig. 1.2). This means that, besides a few exceptions, such as reverse transcription of the RNA genome of retroviruses, genomic DNA stores the building plan of all pro- and eukaryotic organisms. Accordingly, genes are defined as regions of genomic DNA that can be transcribed into RNA. In the original formulation of the dogma only mRNA was meant, i.e. the RNA template used for protein translation, but it applies also to non-coding RNA (ncRNA), such as rRNA, tRNA and micro RNA (miRNA). Nevertheless, the expression of the 20,000 protein-coding genes of the human genome, i.e. their transcription into mRNA and the following translation into protein, determines which proteins are found in a given cell.

An average human cell has only some 100,000 accessible chromatin sites, i.e. some 99% of the genome is not available to DNA binding proteins. However, many of the accessible chromatin regions are not static but dynamically controlled by chromatin modifying and remodeling proteins (Fig. 1.1b, Chap. 6). Some of these proteins are enzymes (referred to as "writers"), such as histone lysine acetyltransferases (KATs), histone lysine methyltransferases (KMTs) and DNA methyltransferases (DNMTs), that add acetyl- or methyl groups to histone proteins or cytosines of genomic DNA, respectively.

These chromatin marks form a "code of instructions" (Chap. 5) that is interpreted by a distinct class of proteins ("readers"), such as methyl-CpG binding domain (MBD) proteins. Almost all chromatin marks can be removed by another class of enzymes ("erasers"), which are lysine deacetylases (KDACs), lysine demethylases (KDMs) and members the ten-eleven translocation (TET) family of DNA demethylases.

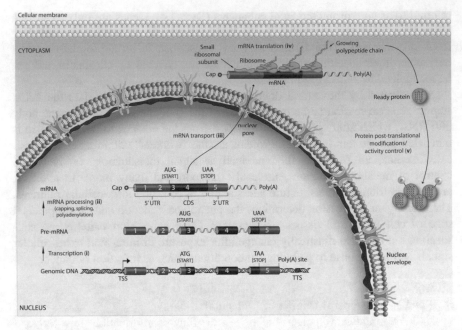

Fig. 1.2 Flow of information from DNA to RNA. The TSS of a gene is the first nucleotide that is transcribed into mRNA, i.e. it defines the 5′-end (the "start") of a gene. In analogy, the 3′-end of a gene is the position where RNA polymerases dissociate from the genomic DNA template. The gene body is entirely transcribed into pre-mRNA, which is composed of exons (numbered *green* and *brown* cylinders) and intervening introns (I). The introns are removed by splicing and the 5′- and 3′-end are protected against digestion by exonucleases through a nucleotide cap and the addition of hundreds of adenines (polyadenylation (poly(A)), respectively (II). Mature mRNA is then exported by an active, i.e. ATP consuming, process through nuclear pores into the cytoplasm (III). There small ribosome subunits scan the mRNA molecules from their 5′-end for the first available AUG (the "start codon"), assemble then with their large subunits and perform protein translation process until they reach UAA, UAG or UGA (the "stop codons") (IV). The mRNA sequence up-stream of the start codon and down-stream of the stop codon are not translated and referred to as 5′- and 3′-untranslated regions (UTRs). Speaking in numbers, the 20,000 human protein-coding genes have an average pre-mRNA length of more than 16,000 nucleotides (nt), while the average human protein is composed of 460 amino acids, for which only 1,380 nt of mature mRNA are needed. This means that only a minor proportion of a gene sequence (some 5–10%, representing only approximately 1% of the human genome) are finally used for coding proteins. The resulting polypeptide chains fold into proteins, most of which are further post-translationally modified, in order to reach their full functional profile (V). Please note that for simplicity in this and in all following figures the nuclear envelope is drawn as single lipid bilayer and not as a double lipid bilayer

In this way, the epigenome instructs the unique gene expression patterns in each of the some 400 tissues and cell types of the human body. Thus, "epigenomics" is the global, comprehensive view of processes that modulate gene expression patterns in a cell independent from genome sequence. These are primarily DNA methylation states (Chap. 4) and covalent modification of histone proteins (Chap. 5) that organize the nuclear architecture, restrict or facilitate

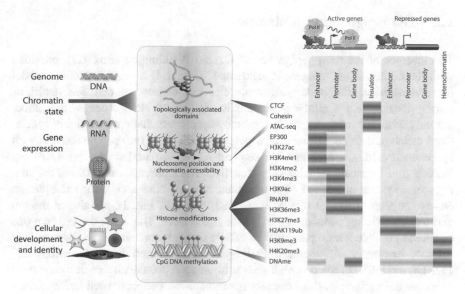

Fig. 1.3 Impact of the epigenome for gene expression. Chromatin acts as a filter for the genome concerning gene expression and in this way determines cell identity (*left*). Epigenomic regulation happens at various scales of chromatin states (*center*), such as topological organization, chromatin accessibility, histone modifications and DNA methylation (Sect. 1.3 and Chap. 3). Key histone modifications and binding nuclear proteins that are characteristic for these chromatin states are indicated and distinguished between active and repressed genes (*right*). The histone modification nomenclature is explained in Box 3.1. CCCTC-binding factor (CTCF) and cohesin are involved in chromatin organization, the KAT EP300 (also called KAT3B) marks enhancers, and both Pol II and H3K36me3 indicate actively transcribed genes. Lighter shades mark a lower or variable degree of protein binding or histone modification, while for DNA methylation it indicates that the sequence could be either methylated or unmethylated

transcription factor access to genomic DNA and preserves a memory of past gene regulatory activities. The epigenome can be considered as the "second dimension" of the genome, which maintains cell-type specific gene expression patterns in normal processes, such as embryogenesis (Chap. 8), as well as in cases of disease, like in cancer (Chap. 10).

During human embryogenesis the fertilized egg, i.e. a single cell, may give rise to all the different tissues and cell types (Sect. 8.1). The different morphologies and functions of these cell types are based on the differential expression of the approximately 20,000 protein-coding genes (and approximately the same number of genes for ncRNA) forming the human genome. This cell type-specific use of the genome is mediated by gene expression programs that are mediated on the action of a subset (in average some 400) of the 1,600 transcription factors encoded by the human genome. Transcription factors can only regulate those genes that have their enhancer and TSS regions within accessible chromatin. Thus, chromatin can be interpreted as a cell type-specific filter of genomic sequence information that determines which genes are transcribed into RNA (Fig. 1.3), in order to ensure a stable differentiated state as well as its inheritance during mitosis.

1.2   The Epigenetic Landscape

The inventor of the term "epigenetics," Conrad Waddington (Box 1.2), provided with the notion of an "epigenetic landscape" (Fig. 1.4) a very illustrative model for understanding the underlying molecular mechanisms of cell fate decisions during development. Cellular differentiation is defined along lineages and under natural conditions represents a forward-moving process that ends in highly specialized cell types. Even in times of a genome-wide view on epigenetics, i.e. with epigenomic methods, such as ChIP-seq and ATAC-seq, and single-cell RNA-seq, Waddington's concept remains very attractive. The landscape model takes the analogy to a system of valleys of a mountain range, where a cell, for example, an embryonic stem (ES) cell (often represented by a ball, Fig. 1.5) begins at the top and follows existing paths driven by gravitational force. This leads the cell into one of several possible fates represented as valleys that get more narrow in the trajectory towards terminally differentiated cell types (also referred to as "canalization," Fig. 1.4, top left). On the downhill path at bifurcation points cell fate decisions need to be taken. In principle, these decisions are stochastic, but once a cell makes a decision, it is restricted in its subsequent decisions by the route it has taken.

Box 1.2 History and Definition of Epigenetics

The discovery of nucleic acids, chromatin and histone proteins in the years 1869 to 1928 allowed the cytological distinction of regions of the cell nucleus into euchromatin (lightly packed) and heterochromatin (tightly packed). In parallel, observations on gene activation via juxtaposition with heterochromatin in fruit flies and transposable elements in maize, respectively, provided hints for – so far not known - non-Mendelian inheritance. Furthermore, the phenomena of X-chromosome inactivation and imprinting (Sect. 4.4) suggested that identical genetic material in the same nucleus can be in an "on" as well as in an "off" state. Based on these observations Conrad Waddington proposed in 1942 on the basis of "epigenesis" (i.e. morphogenesis and development of an organism) the term "epigenetics" for changes in the phenotype without changes in the genotype. Later, this definition was extended to "epigenetics is the study of changes in gene function that are mitotically and/or meiotically heritable and that do not entail change in DNA sequence." This means that epigenetics refers to functionally relevant modifications of the genome, such as DNA methylation and histone modification, without involving a change in the nucleotide sequence. These changes may remain through all the following cell divisions for the remainder of the cell's life and can even last for multiple generations (Sect. 9.2). This implies that epigenetics inherits gene expression patterns by adapting chromatin accessibility and architecture.

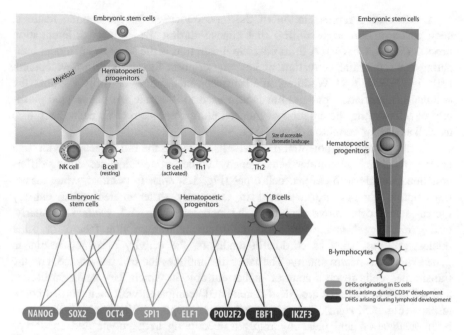

Fig. 1.4 Epigenetic landscape models. Cellular differentiation is accompanied by the progressive restriction of the epigenetic landscape indicated by (i) narrowing of valley floors (*top left*), (ii) changes in the differentiation of state-specific sites of accessible chromatin (*right*) or (iii) the sequential activation of key transcription factors (*bottom left*)

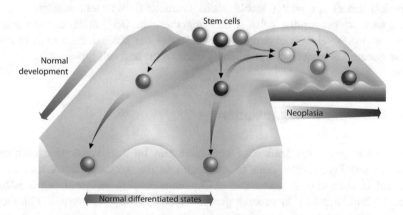

Fig. 1.5 Phenotypic plasticity during cellular reprograming and neoplasia. Waddington's landscape model can be used for the illustration of phenotypic plasticity of cells during normal development (*left*), creation of iPS cells, i.e. in cellular reprograming (*center*), as well as during the induction of neoplasias, i.e. in tumorigenesis (*right*)

An alternative representation of the model focuses on epigenomic features, such as chromatin accessibility, that change during the cellular differentiation process (Fig. 1.4, *right*). A third possible illustration is the "guy-rope" model indicating the sequential activation of key transcription factors during development (Fig. 1.4, *bottom left*). Waddington's landscape model does not only visualize the cellular differentiation process but is also used to illustrate the phenotypic plasticity of cells during the creation of induced pluripotent stem (iPS) cells (Sect. 8.3) as well as during neoplasia (Sect. 10.2).

The developmental potential of stem cells on top of the hill correlates with high entropy mediated by cellular heterogeneity, while entropy declines during differentiation towards well-defined cell types (Fig. 1.5, *left*). In contrast, when master transcription factors are activated in terminally differentiated cells, entropy increases and cells move up-hill in the landscape (Fig. 1.5, *center*). Similarly, during tumorigenesis (epi)mutations can activate transcription factors or other nuclear proteins, such as chromatin modifiers, the activity of which results in gene expression heterogeneity. The latter discontinues the cell fate choice, and the transformed cells reach a state of higher entropy, in which they again proliferate and self-renew, i.e. they are "de-differentiated" compared with their normal counterparts (Fig. 1.5, *right*). Taken together, the concepts of Waddington's model on both canalization and plasticity imply a decoupling of genotype and phenotype and suggest that regulatory processes exist between the two.

A mathematical interpretation of Waddington's landscape identifies the valleys as "attractors" in a multi-dimensional dynamical systems framework. Changes in the epigenome during cellular differentiation result in alternations of the transcriptome. In the mathematical formalism, gene expression profiles are projected onto an n-dimensional phase space of vectors (each representing the expression of an individual gene), in which stable states (attractors) represent intermediate cell types, such as progenitor cells. Taken together, the epigenetic landscape is an attractive, intuitively understandable model how the static information provided by the genome is translated dynamically into tissues and cell types.

1.3   Chromatin Organization

The structure and organization of chromatin can be interpreted as a number of superimposed layers. The core of chromatin is the genomic DNA that can be modified at cytosines, in particular at CG dinucleotides, so-called "CpG islands" (Sect. 4.1 and Box 4.1). In general, genomic DNA is wrapped every 200 bp around histone octamers forming the primary structure of chromatin, i.e. regularly arranged nucleosomes (Sect. 3.1). The nucleosomal histones H2A, H2B, H3 and H4 can be chemically modified and exchanged with histone variants (Sect. 3.3). Furthermore, there are higher order structures of chromatin, in which (i) the chromatin fiber is wrapped around its axis forming a 30 nm fiber, (ii) topologically associated domains (TADs) and (iii) lamin-associated domains (LADs) (Sects. 3.5 and 3.6).

Chromatin, in general, can be distinguished in two main forms: its active form, euchromatin, and its inactive form, heterochromatin. In the latter, CpG islands are methylated, nucleosomes are arranged in a regular dense fashion and histone proteins are tri-methylated at positions H3K9 and H3K27, in order to attract specialized proteins, such as heterochromatin protein 1 (HP1), MBD proteins and Polycomb group (PcG) family members that further seal the chromatin in heterochromatin (Sect. 3.5). Moreover, the respective chromatin regions are located in LADs close to the nuclear envelope. All five layers of chromatin organization (Fig. 1.6, *left*) ensure that genes close to these features stay inactive.

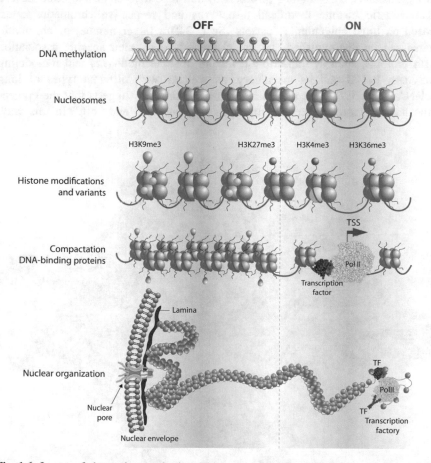

Fig. 1.6 Layers of chromatin organization. There are at least five different layers of chromatin organization that are associated with inactive (off, *left*) or active (on, *right*) transcription. Level 1: Methylated versus unmethylated genomic DNA; level 2: regular nucleosome arrangement versus nucleosome-free regions; level 3: histone tri-methylation at positions H3K9 and H3K27 versus positions H3K4 and H3K36; level 4: densely nucleosome packaging in heterochromatin versus loosely arrangement with transcription factors and Pol II binding in euchromatin; level 5: location within LADs close to the border of the nucleus versus at transcription factories in the center of the nucleus

   In order to activate a gene, the chromatin at its TSS and at enhancer region(s) that control the gene's activity need to be opened, i.e. there has to be a transition from heterochromatin to euchromatin (Chaps. 3 and 4). In brief, this is achieved via the demethylation of CpG islands, evicting individual nucleosomes at TSS regions of housekeeping genes, histone tri-methylation at positions H3K4 and H3K36, the binding of transcription factors and the basal transcriptional machinery including Pol II and the location of the respective chromatin regions close to transcription factories in the center of the nucleus (Fig. 1.6, *right*).

   Genomic regions that are defined via common DNA sequence features, such as CpG islands, TSS regions (promoters) and repetitive elements, often have a characteristic histone modification patterns and respective chromatin states. Based on huge epigenomic datasets (Sect. 2.2) a larger number of chromatin states can be distinguished and are characterized by their specific association with proteins, such as transcription factors, histone modifying and remodeling enzymes, Pol II and its machinery, as well as with different types of long ncRNAs (Fig. 1.7). This more fine-grained distinction suggests that the epigenome has far more differential functions than an "on" and "off." In this way,

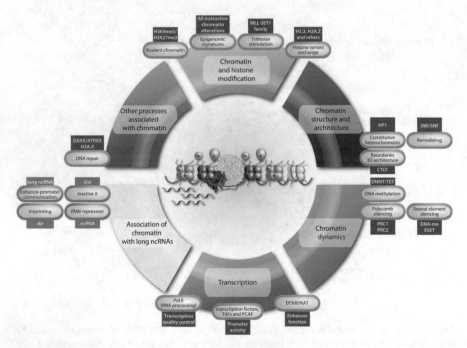

Fig. 1.7 Epigenome function is based on different chromatin states. The different states of chromatin are defined in an inter-dependent fashion by histone modifications, DNA methylation, histone variants and nucleosome remodeling (*center*). Functions of the epigenome are exemplified by 18 chromatin states that are characterized by the specific association with proteins or long ncRNAs

chromatin states do not only characterize genomic elements, such as enhancers, insulators and gene bodies, but also reflect their biochemical activities within the same TAD (Sects. 1.4 and 7.3). TAD boundaries are determined by CTCF binding sites and generally isolate genomic "neighborhoods" from each other (Sect. 3.5).

1.4   Epigenome Marks of Transcriptional Regulation

The TSS region, which is also known as the core promoter, is sufficient to assemble the basal transcriptional machinery including Pol II. However, transcription is often weak in the absence of stimulatory transcription factors. Genomic regions that contain binding sites for sequence-specific transcription factors are called enhancers (Fig. 1.8 *top*). These transcription factors recruit co-activator and co-repressor proteins and the combined regulatory cues of all bound factors determine the activity of enhancers. Enhancers functions via the cooperative binding of multiple proteins in less than one nucleosome length, so that they overcome the energetic barrier of nucleosome eviction (Sect. 7.2). Therefore, enhancer activity is determined by epigenome stages, such as being devoid of nucleosomes, which is often recognized by histone markers of accessible chromatin, such as H3K4me1 and H3K27ac. When enhancers are close (+/- 100 bp) to the TSS, they are also often referred to as promoters, but besides their distance relative to the TSS there are no differences between the terms "enhancer" and "promoter".

Enhancers that regulate the activity of a gene should be located within the same TAD. Since TADs can be as large as 1 Mb, this may be the maximal linear distance between an enhancer and the TSS region(s) that it regulates (Fig. 1.8 *top*). DNA looping events mediated by complexes of the proteins cohesin and CTCF bring transcription factors binding to enhancers into close vicinity of TSS regions, so that they can contact and activate via the Mediator complex the basal transcriptional machinery. The looping mechanism also implies that enhancer regions are as likely upstream as downstream of TSS regions and may have tissue specific usage and effects for transcription, for example, in tissue A enhancer A is used for activation whereas in tissue B for repression (Fig. 1.8 *center and bottom*). Results of large-scale epigenomics projects, such as ENCyclopedia Of DNA Elements (ENCODE, Sect. 2.2), demonstrated that basically all regulatory proteins have a Gaussian-type distribution pattern in relation to TSS regions, i.e. the probability to find an active transcription factor binding site symmetrically declines both up- and downstream of the TSS. Thus, the classical definition of a promoter as a sequence being located only upstream of the TSS is outdated.

Transcription factors are referred to as trans-acting factors, since they are not encoded by the same genomic regions that they are controlling. Accordingly, the process of transcriptional regulation by transcription factors is often called

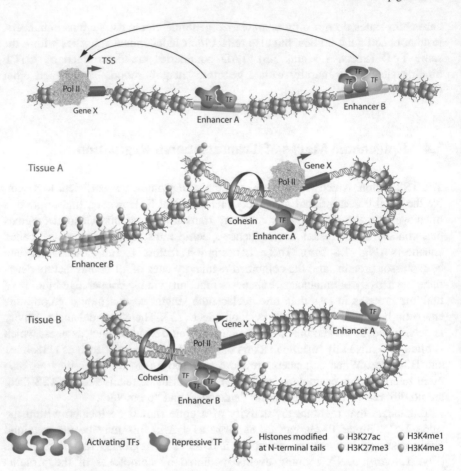

Fig. 1.8 Enhancer function. Enhancers are stretches of genomic DNA that contain binding sites for one or multiple transcription factors (TFs) stimulating the activity of the basal transcriptional machinery bound to the TSS of a target gene. Enhancers are located both upstream and downstream of their target genes in linear distances of up to 1 Mb (*top*). Transcription factor-bound, active enhancers are brought into proximity of TSSs by DNA looping, which is mediated by a complex of cohesin, CTCF and other proteins. Active TSS regions and enhancers show depletion of nucleosomes, while nucleosomes flanking active enhancers have specific histone modifications, such as H3K27ac and H3K4me1 (*center*). In contrast, inactive enhancers are silenced by a number of mechanisms, such as PcG proteins binding to H3K27me3 marks or by binding of repressive transcription factors (*bottom*)

trans-activation. In contrast, enhancers and promoters are clusters of transcription factor binding sites that regulate a gene on the same chromosome, i.e. in *cis*, are often referred to as *cis*-regulatory modules. On the genome scale the complete set of *cis*-regulatory modules is called the cistrome.

TADs often contain multiple genes that are regulated in a common fashion by a special distal enhancer region, called locus control region (LCR) or

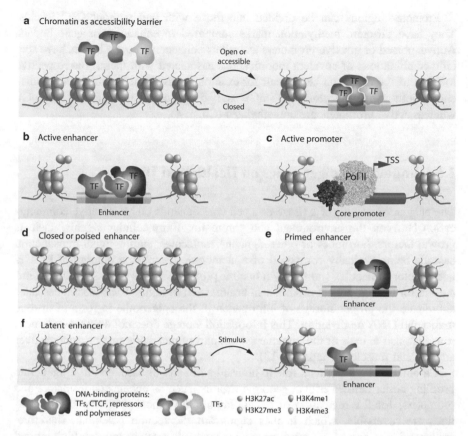

Fig. 1.9 Chromatin accessibility and histone marks. Chromatin restricts the access of transcription factors, Pol II and other nuclear proteins to the genome (a). Enhancers are often marked by H3K27ac and H3K4me1 modifications (b), while H3K27ac and H3K4me3 mark active promoters (c). Closed or poised enhancers carry both active H3K4me1 marks and repressive H3K27me3 marks (d). Primed enhancers are pre-marked by H3K4me1 (e), while latent enhancers with in closed chromatin do not show any specific histone mark, but become accessible through a stimulus (mostly an extra-cellular signal activating a signal transduction cascade) so that flanking nucleosomes acquire H3K4me1 and H3K27ac marks (f)

super-enhancer. Genome-wide there are only a few hundred super-enhancers being characterized by high levels of active histone marks (for example, H3K27ac), strong enrichment with transcription factors and intensive contacts with the Mediator complex. Super-enhancers often are critical in determining the fate of a cell during embryogenesis or differentiation of hematopoietic cell types (Sect. 8.4). At highly transcribed genomic regions, histones in the canonical H2A-H2B/H3-H4 nucleosome are often replaced by their variant forms, such as H2A.Z and H3.3, that locally alter chromatin accessibility (Fig. 1.9) and increase transcription (Sect. 3.3).

Promoter regions can be divided into those with and without CpG islands. They have different methylation marks compared to enhancers or gene bodies. Active, poised or inactive promoters as well as enhancers or gene bodies have specific combinations of covalent modifications associated with them. The respective location of the mark can be critical: for example, actively transcribed gene bodies carry both 5-methylcytosine (5mC) and 5-hydroxymethylcytosine (5hmC), whereas active promoters are unmethylated (Chap. 4).

1.5 Impact of Epigenomics on Health and Disease

The epigenomic status of a tissue or a cell type depends on an effective communication between the environment and chromatin. Extra-cellular signals, such as growth factors, hormones or other signaling molecules, start a signal transduction cascade being typically composed of a membrane receptor, a number of kinases and adaptor molecules and finally a nuclear protein, such as a transcription factor, a co-factor or a chromatin modifier. In health, such signaling transduction cascades coordinate the proper storage of information in the epigenome in form of histone marks and DNA methylation. This information storage does not only apply to neurons, but also to cells of the immune system and of metabolic organs, such as liver and skeletal muscles (Chaps. 11–13).

Chromatin plays a critical role in human health and disease, and its epigenomic profiling better defines critical genomic regions, such as enhancers, insulators and promoters, for the regulation of genes. Importantly, most epigenomic modifications are reversible, which implies significant therapeutic potential. Therefore, epigenomics is one of the most innovative research areas in modern biology and biomedicine where the molecular hallmarks of epigenetic control are used as targets for medical interventions and treatments (Fig. 1.10). For example, iPS cells have the potential to regenerate damaged tissues: they originate from differentiated adult cells, which were overexpressed with master transcription factors binding to super-enhancers, however, their epigenome differs from ES cells. The latter cells are more depleted with respect to marks of repressive chromatin and are more responsive to chromatin remodeling (Sect. 8.3).

Interestingly, during aging the histone acetylation and methylation of many genomic regions changes, most likely because a class of KDACs, called sirtuins (SIRTs), can promote gene silencing and longevity (Sect. 9.3). Similarly, the epigenome of cancer cells is also reprogramed during the transformation process from normal cells. The mapping of active and repressed chromatin regions in cancer cells allows more accurate prognosis and even may facilitate therapy (Sect. 10.5). For example, inhibitors of chromatin modifiers, such as KDAC inhibitors, have been recently approved for cancer treatment. In addition, numerous psychiatric disorders, such as anxiety and depression, can be treated with KDAC inhibitors (Sect. 11.3). Although some treatments with epigenetic inhibitors convincingly ameliorate disease conditions, many compounds are still at an exploratory stage.

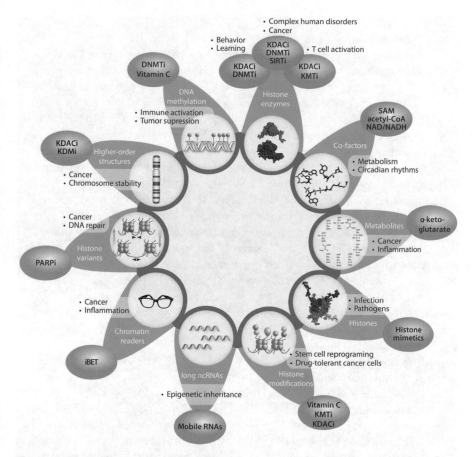

Fig. 1.10 Prognostic and therapeutic potential of epigenomics. Examples for the impact of epigenomics in normal development and in disease are indicated. The circles each represent key epigenomic mechanisms and the mainly associated nuclear proteins. Dys-regulated epigenomics may be reversed by pharmacological intervention with small-molecule inhibitors, such as KDAC inhibitors, DNA methylation inhibitors, SIRT inhibitors, metabolic co-factors, such as S-adenosylmethionine (SAM) and α-ketoglutarate (α-KG), histone lysine methyltransferase inhibitors (KMTi), bromodomain and extra-terminal inhibitors (iBET), poly(ADP-ribose) polymerase inhibitors, histone lysine demethylase inhibitors (KDMi)

Key Concepts
- Epigenetics is defined as the study of the epigenome, and changes in gene functions that are heritable but do not involve changes in the genome.
- The genome-wide analysis of epigenetics is referred to as epigenomics.
- The epigenome is very dynamic, varies from one cell type to the other and can respond to various signaling pathways.

(continued)

Key Concepts (continued)

- Chromatin acts as a filter for the access of DNA binding proteins to functional elements on the genome, such as TSS regions and enhancers.
- The epigenome instructs the unique gene expression patterns in each of the some 400 tissues and cell types of the human body.
- The epigenetic landscape is an attractive, intuitively understandable model how the static information provided by the genome is translated dynamically into tissues and cell types.
- In order to activate a gene there has to be a transition from heterochromatin to euchromatin. This is achieved via the demethylation of CpG islands, evicting individual nucleosome at TSS regions of housekeeping genes, histone tri-methylation at positions H3K4 and H3K36, the binding of transcription factors and the basal transcriptional machinery including Pol II, and the location of the respective chromatin regions close to transcription factories in the center of the nucleus.
- TAD boundaries are determined by CTCF binding sites and generally isolate genomic neighborhoods from each other.
- Enhancers act via the cooperative binding of multiple proteins in less than one nucleosome length, in order to overcome the energetic barrier of nucleosome eviction.
- Basically all regulatory proteins have a Gaussian-type distribution pattern in relation to TSS regions.
- Promoter regions can be divided into those with and without CpG islands and have different methylation marks compared to enhancers or gene bodies.
- The epigenomic status of a tissue or a cell type depends on an effective communication between the environment and chromatin.
- Signaling transduction cascades coordinate the proper storage of information in the epigenome in form of histone marks and DNA methylation.
- Most epigenomic modifications are reversible, which implies significant therapeutic potential. For example, inhibitors of chromatin modifiers, such as KDAC inhibitors, have been recently approved for cancer treatment.

Additional Reading

Allis CD, Jenuwein T (2016) The molecular hallmarks of epigenetic control. Nat Rev Genet 17:487–500

Deans C, Maggert KA (2015) What do you mean, "epigenetic"? Genetics 199:887–896

Perino M, Veenstra GJ (2016) Chromatin control of developmental dynamics and plasticity. Dev Cell 38:610–620

Shlyueva D, Stampfel G, Stark A (2014) Transcriptional enhancers: from properties to genome-wide predictions. Nat Rev Genet 15:272–286

Stergachis AB, Neph S, Reynolds A et al (2013) Developmental fate and cellular maturity encoded in human regulatory DNA landscapes. Cell 154:888–903

Chapter 2
Methods and Applications of Epigenomics

Abstract During the last 10 years the field of epigenomics exploded due to the invention and application of a large number of next-generation sequencing methods investigating various aspects of chromatin biology, such as DNA methylation, histone modification state and 3D structure. Individual research teams as well as large consortia already produced thousands of epigenome maps from hundreds of human tissues and cell types. The integration of these data, for example, transcription factor binding and characteristic histone modifications, allows the prediction of enhancers and promoters as well as monitoring their activity and many additional functional aspects of the epigenome. The continuous development of even more sensitive epigenomic methods, such as ATAC-seq and ChIPmentation, and single-cell approaches in combination with (epi)genetic editing methods, such as CRISPR/Cas9, will allow even more powerful analyses of chromatin states and their associated regulatory networks.

In this chapter, we will compare the current epigenomic methods for the analysis of (i) DNA methylation (bisulfite sequencing), (ii) transcription factor binding and histone modifications (ChIP-seq), (iii) chromatin accessibility (FAIRE-seq and ATAC-seq) and (iv) nuclear architecture (3C, 5C, Hi-C and ChIA-PET). We will present the different epigenomic consortia, such as ENCODE, Functional annotation of the mammalian genome (FANTOM) 5, Roadmap Epigenomics and International Human Epigenome Consortium (IHEC), that apply these methods for high-throughput epigenomic profiling of human cell lines and primary tissues. In this context we will demonstrate that the integration of the different level of data allows inferring function and mechanistic insight, such as delineation of gene regulatory sequences. Finally, we will discuss, how the application of the various epigenomic methods will provide answers to emerging questions in human health and disease.

Keywords Epigenome profile · ChIP-seq · FAIRE-seq · ATAC-seq · 3C · 5C · Hi-C · ChIA-PET · ENCODE · FANTOM5 · Roadmap Epigenomics · IHEC · large-scale data integration · CRISPR/Cas9

© Springer Nature Singapore Pte Ltd. 2018
C. Carlberg, F. Molnár, *Human Epigenomics*, https://doi.org/10.1007/978-981-10-7614-5_2

2.1   Epigenomic Methods

Since more than a decade it is very popular to add the suffix "omics" to a molecular term, in order to express that a set of molecules is investigated on a comprehensive and/or global level. After completing the sequence of the human genome in 2001, "genomics" became the first omics discipline focusing on the study of entire genomes in contrast to "genetics" that investigates individual genes. Genomics turned out to be the appropriate approach for the description and study of genetic variants contributing to complex diseases, such as cancer, diabetes and Alzheimer disease. Technological advances, in particular in next-generation sequencing methods, led to the development of additional omics disciplines, such as epigenomics, transcriptomics, proteomics, metabolomics and microbiomics (Fig. 2.1).

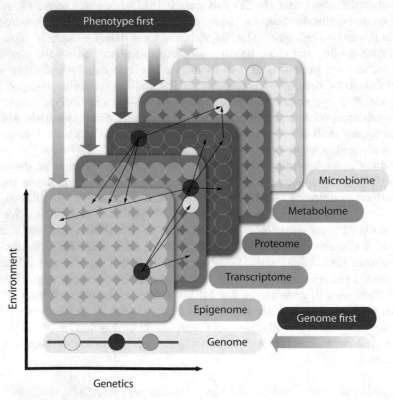

Fig. 2.1 Different layers of omics data types. For the investigation of health and disease omics data can be collected on the indicated levels: genome, epigenome, transcriptome, proteome, metabolome and microbiome. Circles represent individual molecules (genes, RNAs, proteins ...) of these large datasets. Arrows represent potential interactions or correlations detected between molecules of the different layers. The "genome first" approach indicates a start from an associated locus in the genome, while in the "phenotype first" approach any other layer might be the starting point

This allowed investigating a biological or medical question on multiple layers. With the exception of cancer, the genome of an individual is constant over lifetime and identical in all tissues and cell types. This makes genomics attractive as a starting point of biomedical studies. In contrast, all other omics layers are more complex, since they involve cell type-specific differences and are responsive to rapidly changing cellular perturbing signals, such a hormones, cytokines and nutritional molecules. This applies also to epigenomics, which naturally is the layer first in line following genomics. Over the last years deep sequencing machines became the "microscope" in epigenomics research, i.e. in particular this omics discipline rapidly developed due to invention of novel genome-wide methods, the most important of which will be presented in the following.

Maps of human epigenomes can be obtained by technology advancement in applying next-generation sequencing with new biochemical techniques or modifications of established methods, such as chromatin immunoprecipitation (ChIP). Next-generation sequencing methods have the advantage that they provide in unbiased and comprehensive fashion information on the entire epigenome. Thus, conclusions drawn from a few isolated genomic regions may be extended to other parts of the genome.

Global epigenomic profiling allows hypothesis-free exploration of new observations and correlations. In the following the key epigenomic methods for determining (i) DNA methylation, (ii) transcription factor binding and histone modification, (iii) accessible chromatin and (iv) 3D chromatin architecture will be presented. The biochemical core of these methods are: (i) different chemical susceptibility of nucleotides, such as bisulfite treatment of genomic DNA, in order to distinguish between cytosine and 5-methylcytosine (5mC), (ii) affinity of specific antibodies for chromatin-associated proteins, such as transcription factors, modified histones and chromatin modifiers, (iii) endonuclease-susceptibility of genomic DNA within open chromatin compared to inert closed chromatin, (iv) physical separation of protein-associated genomic DNA (of fragmented closed chromatin) in an organic phase from free DNA (of accessible chromatin) in the aqueous phase and (v) proximity ligation of genomic DNA fragments that via looping got into close physical contact.

2.1.1 Mapping DNA Methylation

Cytosine is by far the most dynamic base of the human genome and often becomes methylated at position 5 (5mC, 60–80% of the 28 million CpGs are methylated, Sect. 4.2). Global DNA methylation methods measure cytosine methylation at base resolution over the whole human genome. There are three main methods to measure DNA methylation: (i) digestion of genomic DNA with methyl-sensitive restriction enzymes, (ii) affinity-based enrichment of methylated DNA fragments and (iii) chemical conversion of methylated cytosines. The choice of the respective method depends on the required resolution. Endonuclease digestion-based DNA methylation assays (methylation-sensitive restriction enzyme digestion, MRE-seq) have base resolution but are limited by the cutting frequency of the chosen restriction enzyme

(s) and are biased toward the enzyme recognition sequences. Therefore, this method is not any longer the primary choice. Affinity-based assays using an antibody (methylated DNA immunoprecipitation, MeDIP-seq, or MBD-seq) can enrich methylated fragments from sonicated genomic DNA. Like in ChIP-seq (see below) the resolution of these assays is highly dependent on the DNA fragment size and CpG density, i.e. they are more qualitative than quantitative.

In contrast, bisulfite sequencing is a chemical conversion method that directly determines the methylation state of each cytosine of the whole genome (Fig. 2.2).

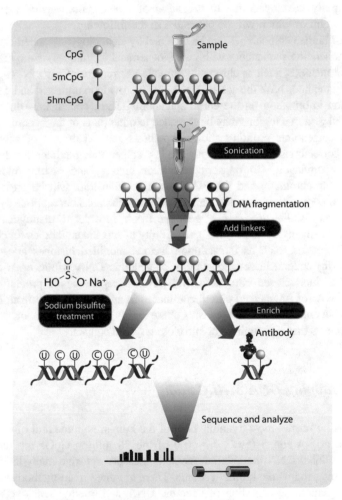

Fig. 2.2 Analysis of DNA methylation. Genome-wide 5-methyl-cytosine (5mC) and its oxidative derivative 5hmC are measured by enrichment- and conversion-based methods followed by massively parallel sequencing. Bisulfite conversion allows quantification of 5mC but does not distinguish 5mC from 5hmC, while antibody enrichment enables qualitative measurement of 5mC and 5hmC. Bisulfite-converted or -enriched genomic DNA is purified, subjected to library construction and clonally sequenced. Bisulfite-converted reads are aligned to the reference genome

Thus, it is mostly used for mapping DNA methylation. Sodium bisulfite treatment of genomic DNA chemically converts unmethylated cytosines to uracils. Through PCR amplification all unmethylated cytosines become thymidines, i.e. remaining cytosines correspond to 5mC. Whole genome sequencing provides then single base resolution of the methylation pattern. However, in addition to 5mC, there are 5-hydroxymethyl-cytosine (5hmC), 5-formyl-cytosine (5fC) and 5-carboxyl-cytosine (5caC). Interestingly, 5mC and 5hmC, but not 5fC or 5caC, are resistant to bisulfite conversion and therefore cannot be distinguished from each other, so that antibody enrichment or more advance chemical conversion methods need to be applied. Bisulfite sequencing is one of the first epigenomic methods being successfully applied on the single-cell level (Box 2.1).

2.1.2 ChIP-seq

ChIP followed by sequencing (ChIP-seq) maps the genome-wide binding pattern of chromatin-associated proteins including post-translationally modified histones. The core of this method is the immunoprecipitation of cross-linked protein-DNA complexes from sonicated chromatin with an antibody that is specific for the protein of interest (Fig. 2.3). Genomic DNA fragments within these complexes are purified from the enriched pool and sequenced. The sequence tags representing the DNA fragments are aligned to the reference genome. Assemblies of the tags are called "peaks" and indicate the genomic regions, where the protein of interest was binding at the moment of cross-linking. ChIP-seq of histone modifications tends to produce broader peaks, i.e. more diffuse regions of enrichment, than transcription factors that bind sequence-specifically and create sharper peak profiles.

> **Box 2.1 Single-cell Analyses**
> Since epigenomic technologies (i) are traditionally performed with larger numbers of cells and (ii) cell populations are known to be heterogenous, the respective results represent average chromatin state for thousands or even millions of cells. However, recent technological advances allowed executing genome-wide analyses on single cells. For example, single-cell RNA-seq showed substantial heterogeneity of cell types in various tissues and identified novel cell populations. The single-cell technology has been extended to the genome and DNA methylome. Bisulfite sequencing of single cells indicates substantial variations in DNA methylation patterns across otherwise homologous cells residing in the same tissues. Recently, single-cell ATAC-seq was developed, which allows single-cell analyses of chromatin accessibility. In general, single-cell epigenomics will provide insights into the combinatorial nature of chromatin, such as which combinations of epigenetic marks and structures are possible and what mechanisms control them.

Although ChIP-seq is a mature method, it is restricted by the need for large amounts of starting material (1–20 million cells), limited resolution and the dependence on the quality of the applied antibodies. Both ENCODE and Roadmap Epigenomics consortia (Sect. 2.2) provide optimized protocols for ChIP-seq and antibody assessment. In the ChIP-exo variation of the ChIP protocol sonicated and immunoprecipitated DNA is treated with a 5′-to-3′ exonuclease, in order to digest DNA to the footprint of the cross-linked protein. The resulting nucleotide resolution protein-binding data allow uncovering motifs of specific binding proteins and the effect of sequence variants on protein-binding affinity. A new variation of ChIP-seq, ChIPmentation, takes advantage of a library preparation using the Tn5 transposase ("tagmentation") as in ATAC-seq (see below). The sequencing library will be prepared using fragmented and immunoprecipitated chromatin, instead of the standard purified, i.e. protein free, immunoprecipitated genomic DNA. This tagmentation step reduces the number of cells needed in the experiments by a factor of 10–100.

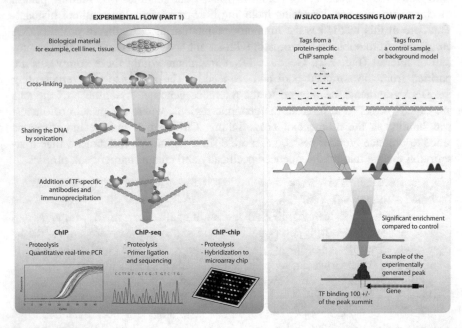

Fig. 2.3 ChIP-seq and its analysis. Short chromatin fragments are prepared from cells, in which nuclear proteins are covalently attached to genomic DNA by short-term formaldehyde cross-linking. Immobilized antibodies against a protein of interest are used to immunoprecipitate the chromatin fragments associated with the respective protein. All genomic fragments are subject for deep sequencing, for example, by the use of an Illumina Genome Analyzer. Typically sequencing runs provide tens of millions of sequencing tags (small arrows) that are uniquely aligned to the reference genome. Clusters of these tags form peaks that represent transcription factor binding loci, when they show significantly higher binding than the control sample

2.1.3 Chromatin Accessibility

Regulatory genomic elements, such as promoters, enhancers and insulators, are bound by sequence-specific transcription factors and associated proteins, such as co-activators and chromatin modifiers. Accessible chromatin is therefore a key characteristic of active genomic regions. The main assays for genome-wide mapping of open chromatin are (i) DNase I hyper-sensitivity followed by sequencing (DNase-seq) and (ii) formaldehyde-assisted identification of regulatory elements followed by sequencing (FAIRE-seq) (Fig. 2.4). The core for the DNase-seq method is a limited digestion of chromatin with the endonuclease DNase I that releases nucleosome-depleted fragments of genomic DNA.

Since nucleosome packing is protecting most regions of genomic DNA against interaction with transcription factors and other nuclear proteins, these regions are also not digestible by DNase I. Sequencing and mapping of these fragments

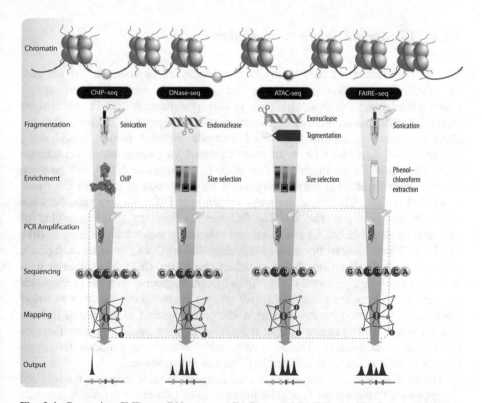

Fig. 2.4 Comparing ChIP-seq, DNase-seq, ATAC-seq and FAIRE-seq. The four genome-wide chromatin profiling methods compared here address different aspects of chromatin structure: ChIP-seq reveals binding sites of transcription factors, modified histones and their associated proteins, while DNase-seq, ATAC-seq and FAIRE-seq report regions of open chromatin. More explanations are provided in the text

identifies DNase I-hyper-sensitive sites that correspond to regulatory genomic regions. High-resolution DNase-seq can even detect transcription factor footprints within larger regions of open chromatin. The FAIRE-seq assay takes advantage of the fact that genomic DNA within open chromatin regions is particularly sensitive to shearing by sonication. Chromatin is isolated from formaldehyde cross-linked cells sonicated and subjected to a phenol-chloroform extraction. Protein-free genomic DNA can be isolated from the aqueous phase, while protein-bound DNA remains in the organic phase. The sequencing and mapping of fragments enriched by FAIRE-seq indicates accessible chromatin.

A rather recently developed method for mapping chromatin accessibility is transposase-accessible chromatin using sequencing (ATAC-seq), which uses Tn5 transposase in the sequencing library preparation (Fig. 2.4). Like in ChIPmentation, the use of tagmentation largely reduces the number of cells needed per experiment.

2.1.4   Chromatin Interaction Analyses

The 3D organization of chromatin affects basically all processes in the nucleus, such as DNA replication, transcription and DNA repair. Fluorescent microscopy imaging methods, such as fluorescence *in situ* hybridization (FISH), can analyze individual cells, but are limited in resolution and throughput. In contrast, chromosome conformation capture (3C)-based methods combine protein cross-linking and proximity ligation of DNA, in order to detect long-range chromatin interactions. They quantify the interaction frequency between genomic loci that are close to each other in 3D, but may be separated by thousands of bases in the linear genome (Fig. 2.5). This identifies loops of genomic DNA, for example, between promoter and enhancer regions. The 3C method involves (i) cross-linking of segments of genomic DNA to proteins and between proteins with each other (like in ChIP), (ii) restriction digestion of the cross-linked DNA, in order to separate non-cross-linked DNA from the cross-linked chromatin, (iii) intra-molecular ligation of neighboring, previously cross-linked DNA fragments with the corresponding junctions ("proximity ligation"), (iv) reverse cross-linking resulting in linear DNA fragment with a central restriction site corresponding to the site of ligation and (v) qPCR using primers against the site of ligation measuring quantitatively the fragment of interest. The frequency with which two genomic fragments become ligated indicates how often they interact in the nucleus.

The main difference between 3C-based methods is their scope. In the original 3C approach ("one versus one") the interaction of two pre-determined genomic regions is studied, chromosome conformation capture-on-chip (4C, "one versus all") captures the interaction of one region with the rest of the genome and chromosome conformation capture carbon copy (5C, "many versus many") uses thousands of anchor and bait primers that span over a sub-region of a chromosome (mostly below 1 Mb). In the genome-wide version of 3C, high-throughput

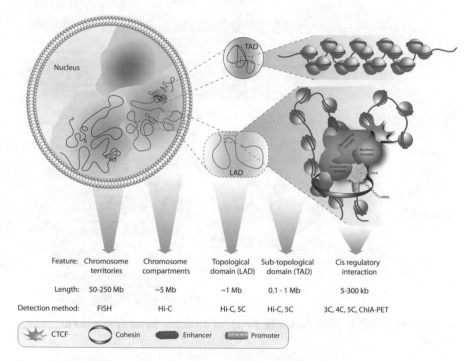

Feature:	Chromosome territories	Chromosome compartments	Topological domain (LAD)	Sub-topological domain (TAD)	Cis regulatory interaction
Length:	50-250 Mb	~5 Mb	~1 Mb	0.1 - 1 Mb	5-300 kb
Detection method:	FISH	Hi-C	Hi-C, 5C	Hi-C, 5C	3C, 4C, 5C, ChIA-PET

CTCF Cohesin Enhancer Promoter

Fig. 2.5 Hierarchical principles of nuclear organization. Details are provided in the text

chromosome capture (Hi-C, "all versus all"), massive parallel sequencing is used. Since only a fraction of DNA fragments generated by 3C-based methods are correct ligation products between distinct genomic loci, ligation junctions are enriched by biotin-labeling and affinity purification. Nevertheless, even for 50 kb-resolution, Hi-C analyses of the human genome require extremely high sequencing depths, and a 1 kb resolution requires approximately 5 billion valid chromatin contacts. Moreover, the method chromatin interaction analysis with paired-end tag sequencing (ChIA-PET) incorporates a ChIP step into the 3C protocol and enriches interactions between genomic regions that are bound by specific proteins. This method achieves a higher resolution than Hi-C, since only ligation products involving the immunoprecipitated molecules are sequenced.

2.2 Epigenomics Consortia

With a delay of some 20 years molecular biologists followed the example, of physicists and realized that some of their research aims can only be reached by international collaborations of dozens to hundreds of research teams and institutions in so-called "Big Biology" projects (Fig. 2.6). The Human Genome Project (www.genome.gov/10001772), which was launched in 1990 and completed in

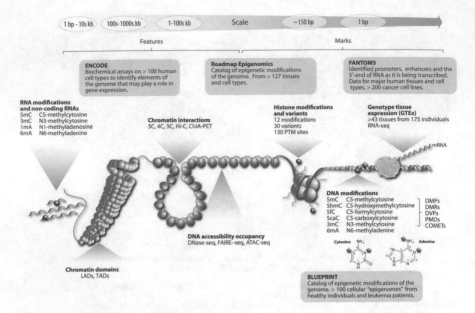

Fig. 2.6 Big Biology projects. Outline of the type of datasets collected by different Big Biology projects

2001, was the first example and has significantly changed the way of thinking in the bioscience community, as a consequence of which more and more single gene studies shifted over to the genome scale.

Since humans are the main intended beneficiaries of medical research, several consortia are focused on the data collection from human tissues and cell types. They have produced a large amount of transcriptomics and epigenomics data in multiple tissues. For example, the Roadmap Epigenomics Project (http://www.roadmapepigenomics.org) and Genotype Tissue Expression (GTEx, http://www.gtexportal.org/home) analyzed epigenomic signatures and transcriptomics in dozens of human tissues and cell types. The first big biology projects in the field of epigenomics and transcriptomics were ENCODE (Encyclopedia of DNA Elements, https://www.encodeproject.org) and FANTOM (http://fantom.gsc.riken.jp). Finally, IHEC (http://ihec-epigenomes.org) became the umbrella organization under which national and international epigenome efforts are jointly coordinated.

2.2.1 ENCODE

ENCODE started in 2003 and was the first international project that used large-scale epigenomic profiling for identifying regulatory elements in the human genome. ENCODE systematically mapped regions of transcription, transcription factor association, chromatin structure and histone modification. Across

147 cell types the project used ChIP-seq for sampling many components of the transcriptional machinery, 119 of the approximately 1,600 transcription factors encoded by the human genome as well as 13 histone or DNA modifications. Moreover, also DNase-seq, FAIRE-seq and RNA-seq analyses have been undertaken on many cell types. However, the project focused on cell lines rather than tissues or primary cells. Subsequently, ENCODE was expanded to include model organisms (modENCODE), in order to add the power of comparative epigenomics. The main results of the ENCODE project are (as reported in *Nature* 489, 57–74):

1. The vast majority (some 80%) of the human genome participates in at least one biochemical RNA- and/or chromatin-associated event in at least one of 147 studied cell types.
2. The human genome contains some 400,000 regions with enhancer-like features and approximately 70,000 regions with promoter-like features.
3. Many novel ncRNAs have been identified both within protein-coding genes as well as in inter-genic regions.
4. Many regulatory clusters, i.e. regions that contain multiple transcription factor binding sites, are located close to a newly identified TSS region. This suggests that many of these regulatory clusters are undiscovered TSS regions and not enhancers.
5. Regulatory sequences are mostly symmetrically distributed around TSS regions, respectively do not show preference towards upstream regions, as was suggested earlier. However, a few histone marks and Pol II signals are clearly asymmetrical, with far higher levels of Pol II in transcribed regions than in upstream regions.
6. The accessibility of chromatin and the patterns of histone modifications are efficiently predicting TSS regions and their activity.
7. Distal regions of open chromatin have characteristic histone modification patterns that distinguish them from TSS regions.
8. Comparative (epi)genomics indicated that some 40% of the constrained sequences of the human genome are located within protein-coding exons and associated untranslated regions, but the majority (60%) is found within the non-coding portion. Some 30% of the latter overlap with experimentally verified non-coding functional regions.

Overall, the project provides new insights into the organization and regulation of the human genome (Sect. 2.3).

2.2.2   FANTOM5

The focus of the FANTOM5 project is on cell-type-specific transcriptomes, but for the annotation of the latter also epigenome profiles were collected. Using the method cap analysis of gene expression (CAGE) the project mapped TSSs and their usage in approximately 750 samples of primary human tissues and cell types in reference to some 250 different human cell lines. CAGE allows separate analysis of multiple promoters linked to the same gene. In this way, FANTOM5 produced a comprehensive map of tissue-specific gene expression across the human body. The major finding of the project were (as reported in *Nature* 507, 462–470):

1. At least one promoter was identified for more than 95% of annotated protein-coding genes in the human genome.
2. Only a few genes showed to be true "housekeeping" genes, i.e. to be always expressed and not responsive to external stimuli.
3. Human promoters are composite entities composed of several closely separated TSSs with independent cell-type-specific expression profiles.
4. TSSs specific to different cell types evolve at different rates, whereas promoters of broadly expressed genes are the most conserved.
5. Promoter-based expression analysis revealed key transcription factors defining cell states.
6. Cancer cell lines generally fail to cluster in a sample to-sample correlation graph with their supposed cell type or tissue of origin and express more transcription factors than primary cells. Thus, the use of primary cells is the logical choice, in order to build mammalian transcriptional regulatory network models that reflect the normal untransformed state.
7. Each primary cell expresses in average some 430 transcription factors at appreciable levels. The ranking of these transcription factors by expression strength can be used to reduce the complexity of the primary cell types and may identify key known regulators of the respective cellular phenotypes.
8. Based on 135 tissue and 241 human cell line samples some 43,000 enhancers were identified.
9. Enhancers share properties with CpG-poor mRNA promoters and produce bidirectional capped RNAs, so-called enhancer RNAs (eRNAs, Box 2.2), that are robust predictors of enhancer activity in a cell.

Box 2.2 eRNAs

eRNAs are ncRNAs that represent Pol II transcripts of enhancer regions. Most eRNAs result from a non-productive scanning of genomic DNA by Pol II, where highly accessible enhancer regions have the highest chance to be transcribed. In addition, DNA looping brings enhancer regions into the proximity of the basal transcriptional machinery residing on TSS regions, which further increases the probability that Pol II initiates transcription at enhancer regions. The number of different eRNAs expressed in human cells is higher than 40,000 and the FANTOM5 project used their mapping for the identification of functional enhancer regions. However, eRNAs may primarily represent transcriptional noise, i.e. in most cases they may not reflect a specific function.

The collection of active enhancers presented by FANTOM5 provides a resource that complements the findings of ENCODE.

2.2.3   Roadmap Epigenomics

The human genome sequence, as created by the Human Genome Project some 15 years ago, is the reference for basically all genomic studies in humans. Similarly, in 2015 the Roadmap Epigenomics Consortium provided human epigenome references from 111 primary human tissues and cell lines. Together with 16 samples already provided by ENCODE, there are 127 reference epigenomes available (Fig. 2.7). This implies that for these epigenomes the complete set of profiles for histone modification patterns, DNA accessibility, DNA methylation and RNA expression were determined. These data allowed establishing global maps of regulatory elements, such as promoters, enhancers and insulators (Sect. 2.3). The main findings of the consortium were (as reported in Nature *518*, 317–330):

1. Histone mark combinations show distinct levels of DNA methylation and accessibility, and predict differences in RNA expression levels that are not reflected in either accessibility or methylation (Sect. 5.3).
2. Megabase-scale regions with distinct epigenomic signatures show strong differences in activity, gene density and nuclear lamina associations.
3. On average two-thirds of each reference epigenome are quiescent.
4. Approximately 5% of each reference epigenome shows enhancer and promoter signatures, which are on average two-fold enriched for evolutionarily conserved non-exonic elements.

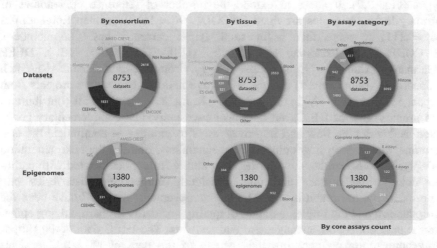

Fig. 2.7 IHEC datasets. 8,753 datasets (June 2017) represent 1,380 different human epigenomes, the majority (68%) of which were obtained from blood. However, only 127 epigenomes, as provided by ENCODE and Roadmap Epigenomics are considered as complete, i.e. the are described by nine or more different epigenomic assays

5. Epigenomic data sets can be imputed at high resolution from existing data, completing missing marks in additional cell types, and providing a more robust signal even for observed data sets.
6. Dynamics of epigenomic marks in their relevant chromatin states allow a data-driven approach to learn biologically meaningful relationships between cell types, tissues and lineages.
7. Enhancers with coordinated activity patterns across tissues are enriched for common gene functions and human phenotypes, suggesting that they represent coordinately regulated modules.
8. Regulatory motifs are enriched in tissue-specific enhancers, enhancer modules and DNA accessibility footprints, providing an important resource for gene-regulatory studies.
9. Genetic variants associated with diverse traits (i.e. cellular or molecular properties, such as body mass index (BMI) or type 2 diabetes (T2D) risk) show epigenomic enrichments in trait-relevant tissues, providing an important resource for understanding the molecular basis of human disease.

Taken together, Roadmap Epigenomics provides the so far the most comprehensive map of the epigenomic landscape of primary human tissues and cell types.

2.2.4 IHEC

IHEC is an overarching epigenomic consortium that was launched in 2010, in order to coordinate the production of human reference epigenomes of major cell types (Fig. 2.7). It seeks to extend the number of reference epigenomes of primary human samples provided ENCODE ($n = 16$), Roadmap Epigenomics ($n = 111$) to more than a thousand. IHEC currently has nine members: ENCODE (US), Roadmap Epigenomics (US), BLUEPRINT (EU), DEEP (Germany), Canadian Epigenetics and Environment and Health Research Consortium (both Canada) and the respective national epigenome projects from Japan, South Korea, Singapore and Hong Kong. The members all contribute to the primary goal of IHEC, but they also have individual complementary goals, such as developing new and improved ways to monitor or manipulate the epigenome, discovering new epigenomic mechanisms, training the next generation of epigenome researchers, exploring epigenomic features associated with pathophysiology and disease and thus translating epigenomic discoveries into improvements to human health. The key achievements of IHEC have been the introduction and the implementation of quality standards for harmonizing epigenomic data collection, management and analysis. The IHEC data portal (http://epigenomesportal.ca/ihec) provides access to the data of all IHEC projects (Fig. 2.7). From all profiled chromatin marks histone modifications represent by far the largest category (Chap. 5).

2.3   Exploring Epigenomics Data

Epigenome profiling leads to maps on DNA methylation, histone marks, DNA accessibility and DNA looping that can be visualized on appropriate web pages, such as the UCSC Genome Browser (Box 2.3). Although this visualization can be highly illustrative and may induce hypotheses, epigenome maps are primarily descriptive (Fig. 2.8), i.e. they are used for annotation. However, enhancers, promoters and other genomic features have characteristic epigenomic signatures, such as H3K4me1 marks for enhancers and H3K4me3 marks for promoters (Chap. 5), on the basis of which they can be identified within epigenome maps. This is comparable to the situation after the sequencing of the human genome, where only 1.5% of it was understood in its function, such as protein coding. Thus, in order to infer function from epigenomics data, they need to be integrated within the same layer (Fig. 2.1), such as histone marks with DNA accessibility, or across layers, such as transcription factor binding with mRNA expression as obtained by the GTEx project or DNA methylation with genetic variation described in the genome-wide association (GWAS) catalog (https://www.ebi.ac.uk/gwas).

Box 2.3 Visualizing Epigenomic Data
A typical way of visualizing epigenomic data, such as those from ENCODE, is to display a selected subset of them in a browser, like the UCSC Genome Browser (http://genome.ucsc.edu/ENCODE). Datasets can be inspected without downloading them by creating a dynamic UCSC Genome Browser track hub that can be visualized on a local mirror of the UCSC Browser. Other visualization tools supporting the track hub format, such as Ensembl, can also be used. For every given genomic position a graphical display provides an intuitively understandable description of chromatin features, such as histone acetylation and methylation, that can be read in combination with experimentally proven information about transcription factor binding, as obtained from ChIP-seq experiments.

Ideally, one of the 127 reference epigenomes provided by ENCODE and Roadmap epigenomics should be included in the analysis. Disease- and trait-associated genetic variants are often show tissue-specific enrichment for enriched in epigenomic marks. This demonstrates the central role of epigenome-wide information for understanding gene regulation as well as embryogenesis and cellular differentiation (Chap. 8). Moreover, the broad coverage of epigenomic annotations improves the understanding of common diseases, such as cancer, Alzheimer disease, autoimmune diseases and diabetes, beyond the level of protein-coding genes (Chaps. 10–13).

Based on IHEC standards each reference epigenome needs to be composed of at least nine profiles and assays, but typically reference epigenomes are composed

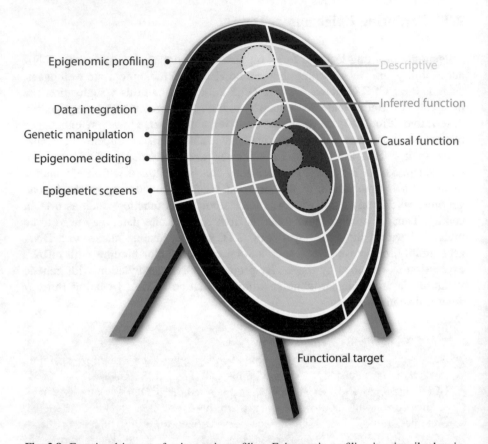

Fig. 2.8 Functional impact of epigenomic profiling. Epigenomic profiling is primarily descriptive, but data integration may allow inferring functions. Genetic manipulation of epigenetic features could reveal, at least indirectly, their functional relevance, while epigenome editing of single epigenomic targets may prove causality. Future approaches, such as high-throughput epigenome editing, should allow the epigenome-wide identification of novel functional marks

of 20–50 genome-wide profiles and represent a multi-dimensional data matrix. There are a variety of approaches to integrate epigenomics data within or across omics layers, such as correlation or co-mapping. When the datasets, which are to be compared, have a common driver or if one regulates the other, correlations or associations should be observed. This requires the application of appropriate statistical methods, many of which have been recently developed for the omics field. In most cases more than two datasets, which often derive from different omics layers, are integrated (often referred as "modeled") in networks (Fig. 2.9). The multi-omics datasets should be collected on the same set of samples, which is not always possible, such as in the case of gene expression and GWAS data.

The most prominent results of the integration of data from ENCODE and Roadmap Epigenomics were already listed in section 2.2. A further illustrative

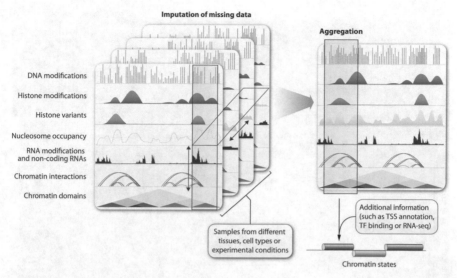

Fig. 2.9 Multi-dimensional integration of epigenome profiles. Epigenome data integration is achieved through (i) imputation of missing data via profiles from the same and/or closely related samples and (ii) addition of non-epigenomic data, such as transcriptomic data (for example, gene expression levels and TSS use). This allows the aggregation and segmentation of the datasets into a number of different chromatin states

example for efficient epigenome integration is the mapping of enhancer modules (describes as accessible genomic regions determined by DNase-seq) over the 111 reference epigenomes of Roadmap Epigenomics (Fig. 2.10). It shows that a limited set of enhancer modules are active in nearly all tissues and cell types, while the majority of the enhancers are specific to same lineage, such as stem cells or cells from the blood. This demonstrates that the activity of enhancer modules can be used to monitor the similarity and relationship of cell types, including common functions and phenotypes. This is useful in the interpretation of human genetic variation and disease, as the enrichment of their traits is strongest for enhancer-associated marks.

Epigenomic profiling facilitates the discovery of the multitude of coordinated chromatin changes that occur, for example, during development (Chap. 8) and disease (Chaps. 10–13). Data integration reduces the list of affected epigenomic regions to a subset with inferred function. For the analysis of the functional impact of presently uncharacterized epigenomic features, individual epigenomic regions or whole epigenome maps, experiments need to be designed that specifically target these sites in the epigenome, such as epigenome editing or epigenetic screens (Fig. 2.8). The targeting of a precise epigenomic location may be most suitable by a fusion of a chromatin modifier with the CRISPR-Cas9 system. Importantly, due to the use of a so-called guide RNA (gRNA) the fusion complex can recognize any position within the genome, even if it is covered by chromatin. Moreover, the CRISPR-Cas9 system allows multiplexing, since only the specific gRNAs need to be exchanged.

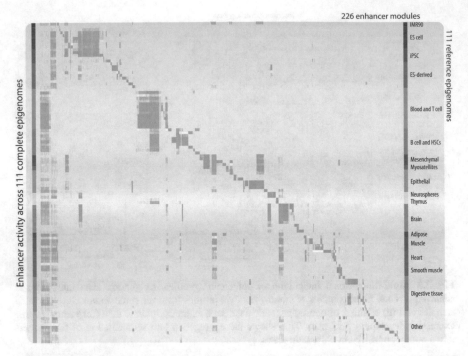

Fig. 2.10 Mapping enhancer modules. Regulatory modules, such as enhancers, can be identified based on activity-based clustering of 2.3 million accessible genomic regions across 111 reference epigenomes (horizontal lines). Vertical lines separate 226 enhancer modules. Data were taken from the Roadmap Epigenomics Consortium (Nature *518*, 317–330)

Although the more than 1,300 epigenomes from hundreds of human tissues and cell types already significantly improved insight into epigenomics, much more remains to be investigated. For example, better characterizations of epigenome variations in human populations and in patients will be critical, in order to fully appreciate the potential of epigenomics in human health and disease.

Key Concepts
- Maps of human epigenomes can be obtained by technology advancement in applying next-generation sequencing with new biochemical techniques or modifications of established methods, such as ChIP.
- Next-generation sequencing methods have the advantage that they provide in unbiased and comprehensive fashion information on the entire epigenome.
- Cytosine is by far the most dynamic base of the human genome and often becomes methylated; 60–80% of the 28 million CpGs are methylated.
- Bisulfite sequencing is one of the first epigenomic methods being successfully applied on the single-cell level.

(continued)

Key Concepts (continued)

- The method ChIP-seq is used to map the genome-wide binding pattern of chromatin-associated proteins including post-translationally modified histones. The core of this method is the immunoprecipitation of cross-linked protein-DNA complexes from sonicated chromatin with an antibody that is specific for the protein of interest.
- The main assays for genome-wide mapping of open chromatin are DNase-seq, FAIRE-seq and ATAC-seq.
- 3C-based methods combine protein cross-linking and proximity ligation of DNA, in order to quantify the interaction frequency between genomic loci that are close to each other in 3D, but may be separated by thousands of bases in the linear genome.
- The first big biology projects in the field of epigenomics and transcriptomics were ENCODE and FANTOM. ENCODE analyzed the epigenome and transcriptome of nearly 150 human cell lines and FANTOM5 used the method CAGE, in order to produce a comprehensive map of tissue-specific gene expression across the human body.
- The projects Roadmap Epigenomics and GTEx analyzed epigenomic signatures and transcriptomics in dozens of human primary tissues and cell types.
- There are 127 human reference epigenomes available.
- IHEC is the umbrella organization under which international epigenome efforts are jointly coordinated.
- The broad coverage of epigenomic annotations improves the understanding of common diseases, such as cancer, Alzheimer disease, autoimmune diseases and diabetes, beyond the level of protein-coding genes.
- The activity of enhancer modules can be used to monitor the similarity and relationship of cell types, including common functions and phenotypes.
- Epigenomic profiling facilitates the discovery of the multitude of coordinated chromatin changes that occur, for example, during development and disease.

Additional Reading

Andersson R, Gebhard C, Miguel-Escalada I et al (2014) An atlas of active enhancers across human cell types and tissues. Nature 507:455–461

Encode-Project-Consortium (2012) An integrated encyclopedia of DNA elements in the human genome. Nature 489:57–74

Fantom-Consortium, Hoen PA, Forrest AR et al (2014) A promoter-level mammalian expression atlas. Nature 507:462–470

Friedman N, Rando OJ (2015) Epigenomics and the structure of the living genome. Genome Res 25:1482–1490

Hasin Y, Seldin M, Lusis A (2017) Multi-omics approaches to disease. Genome Biol 18:83

Meyer CA, Liu XS (2014) Identifying and mitigating bias in next-generation sequencing methods for chromatin biology. Nat Rev Genet 15:709–721

Rivera CM, Ren B (2013) Mapping human epigenomes. Cell 155:39–55

Roadmap Epigenomics Consortium, Kundaje A, Meuleman W et al. (2015) Integrative analysis of 111 reference human epigenomes. Nature 518:317–330

Schmitt AD, Hu M, Ren B (2016) Genome-wide mapping and analysis of chromosome architecture. Nat Rev Mol Cell Biol 17:743–755

Stricker SH, Koferle A, Beck S (2017) From profiles to function in epigenomics. Nat Rev Genet 18:51–66

Part B
Molecular Elements of Epigenomics

Chapter 3
The Structure of Chromatin

Abstract Chromatin is the physical representative of the epigenome. Histones are the main protein component of chromatin and form octamer cores, around which genomic DNA is wrapped. These nucleosomes are the regularly repeating units of chromatin, but they can vary from one genomic region to the other by (i) different post-translational modifications of amino acid residues of the histone proteins and (ii) the introduction of histone variants. These genomic site-specific modifications are reversible and an important component of the epigenomic memory affecting transcription factor binding and differential gene expression in every cell type. Moreover, the spatial organization of chromatin not only allows dense packaging of otherwise "long" genomic DNA into a small nucleus, but is also an important level of distinction of open chromatin (euchromatin) and closed chromatin (heterochromatin). Chromatin architecture is formed in a hierarchical manner and engages dynamic chromatin loops. The formation of these loops is mediated by architectural proteins, such as CTCF and cohesin, and other regulatory components, such as transcription factors, PcG proteins and heterochromatin. This gives rise to structural landmarks, such as TADs and chromosome territories.

In this chapter, we will introduce the different levels of chromatin structure and their functional impact. We will discuss histone proteins, their post-translational modifications and their variants. Moreover, we will present the different regulatory layers of chromatin ranging from single nucleosomes, looping of genomic regions as well as large-scale folding of whole chromosomes into territories within the nucleus.

Keywords Chromatin · histone proteins · nucleosome · histone variants · post-translational histone modifications · DNA looping · CTCF · cohesin · TADs · LADs · chromosome territories

3.1 Central Units of Chromatin: Nucleosomes

Genomic DNA is, due to its phosphate backbone, at physiologic pH negatively charged. The electrostatic repulsion between adjacent DNA regions makes it impossible to fold long DNA molecules of the individual chromosomes into the

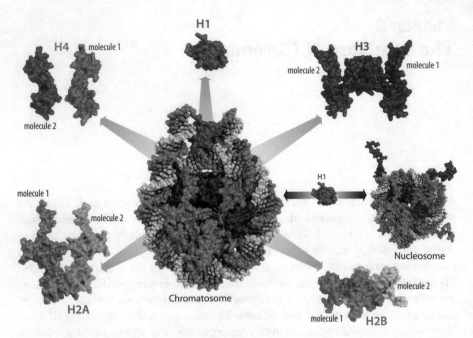

Fig. 3.1 The chromatosome. This space-filling surface representation of a nucleosome contains each two copies of the four core histone proteins H2A (*green*), H2B (*orange*), H3 (*red*) and H4 (*blue*) and 147 bp of genomic DNA (*grey*) being wrapped 1.8-times around the histone core. In complex with the linker histone H1 the nucleosome is also referred to as chromatosome

limited space of the nucleus. Nature solved this problem by wrapping genomic DNA around a complex of positively charged histone proteins. Each two copies of the histone proteins H2A, H2B, H3 and H4 form a histone octamer (Fig. 3.1), which in complex with 147 bp genomic DNA – wrapped nearly twice around it – is referred to as the nucleosome. The bending of DNA is mainly enabled through the attraction between the positively charged histone tails and the negatively charged DNA backbone. In addition, at some genomic regions the bending is supported by natural curvature of DNA that is achieved by AA/TT dinucleotides repeating every 10 bp and a high CG content. Together with the linker histone H1 the nucleosome forms the so-called chromatosome. Each chromatosome is connected by 20–80 bp linker DNA with the following one. This forms a repetitive unit approximately every 200 bp of genomic DNA. This regular positioning of nucleosomes has the effect that the position of one nucleosome determines the location of its neighbors. However, chromatin remodeling complexes have an ATP-driven core that regulates the position and composition of nucleosomes (Chap. 7).

In every cell of the human body the diploid genome is covered by approximately 30 million nucleosomes. This chromatin fiber has a diameter of 10 nm and often is referred to as "beads on a string." The phosphate backbone of 200 bp genomic DNA carries 400 negative charges, which are in part neutralized by the approximately 220 positively charged lysine and arginine residues of the histone octamer. Thus, higher order folding of chromatin requires the neutralization of the remaining (180) negative charges by the linker histone H1 and other positively charged nuclear proteins.

3.2 Histones and Their Modifications

The general feature of the core histone proteins is their small size of some 11–15 kD and their disproportional high content of the basic amino acids lysine and arginine, in particular at their amino-termini. Each of the four core histones H2A, H2B, H3 and H4 comes in multiple variants and more than 100 human genes are coding for histone proteins. This makes the histone gene family one of the largest within the human genome. In contrast to transcription factors, the DNA binding of histones is not sequence specific.

The tails and globular domains of histone proteins provide over 130 sites for post-translational modifications (Box 3.1). The best-characterized covalent modifications on histones are methylation at lysines (K) and arginines (R), acetylation at lysines (K) and phosphorylation at serines (S) and threonines (T), but also ubiquitination, crotonylation, succinylation and malonylation have been observed. In a rough distinction the role of histone modifications is either to activate or to silence gene transcription. Moreover, the histone marks control the accessibility of DNA and serve as docking site for the recruitment (or exclusion) of protein complexes. For example, H3K27ac is found both at active promoters and enhancers, H3K36me3 marks actively transcribed gene bodies and H3K27me3 indicates heterochromatic or repressed regions (Chap. 5).

Box 3.1 Nomenclature of Histone Modifications
Histone modifications are named according to the following rule:

the name of the histone protein (for example, H3).
the single-letter amino acid abbreviation (for example, K for lysine) and the amino acid position in the protein.
the type of modification (ac: acetyl, me: methyl, P: phosphate, Ub: ubiquitin), a complete list is indicated in Fig. 5.1.
the number of modifications (only methylations are known to occur in more than one copy per residue, thus 1, 2 or 3 indicates mono-, di- or tri-methylation).

As an example H3K4me3 denotes the tri-methylation of the fourth residue (a lysine) from the amino-terminus of the protein histone 3. This histone modification serves as a general mark for active promoter regions.

The enzymes responsible for histone modifications ("writers") are often highly residue-specific (Chap. 6). Covalent modifications of histone proteins alter the physio-chemical properties of the nucleosome and are recognized by specific proteins ("readers"). Similarly, basically all covalent histone modifications are reversible via the action of specific enzymes ("erasers"). For example, when a KAT enzyme adds an acetyl group to the amino group of K4 of histone H3, the positive charge of this amino acid is neutralized (Fig. 3.2). In reverse, KDACs can remove the acetyl group from H3K4ac and restore the positive charge of the lysine residue. In principle, lysine acetylation is possible for every accessible residue, i.e. primarily for those that are located within the unstructured histone tail. This example indicates, how chromatin modifiers determine through the addition or removal of a rather small acetyl group the charge of the nucleosome core. This has major impact on the attraction between nucleosomes and the density of chromatin packing. Thus, the primary chromatin fiber is by far not as uniform as initial structure analysis suggested but can be very variable from nucleosome to nucleosome.

3.3   Histones Variants

The "canonical" histones H2A, H2B, H3 and H4 comprise the majority of histone species in any human cell. In addition, there are eight variants of H2A (H2A.X, H2A.Z.1, H2A.Z.2.1, H2A.Z.2.2, H2A.B, macroH2A1.1, macroH2A1.2 and macroH2A2), two variants of H2B (H2BFWT and TSH2B) and six variants of H3 (H3.3, histone H3-like centromeric protein A (CENP-A), H3.1T, H3.5, H3.X and H3.Y). In humans and other higher eukaryotes there are no variants of H4. Most histone variants are expressed in nearly all somatic tissues, but H2BFWT and TSH2B are exclusively found in the male germline.

Canonical histones are assembled into nucleosomes behind the replication fork to package newly synthesized genomic DNA. By contrast, the incorporation of histone variants into chromatin is independent of DNA synthesis and occurs throughout the cell cycle. The genes of canonical histones are organized as clusters of multiple copies. This ensures their coordinated high expression at equal amounts during the S phase of the cell cycle. In contrast, each histone variant is encoded by only one or two gene copies that are distributed over the human genome. This allows their specific gene expression throughout all phases of the cell cycle. Interestingly, canonical histones have no introns, i.e. they have no splice variants, while most of the genes for the histone variants do have introns and alternative spice variants.

Histone variants are often subjected to the same modifications as canonical histones. For example, K4 of H3.3 gets tri-methylated (H3.3K4me3) and K18 and K23 are acetylated (H3.3K18ac and H3.3K23ac, respectively). However, there are also variant-specific modifications on residues that differ from their canonical counterparts. Accordingly, histone variants also directly influence the structure of

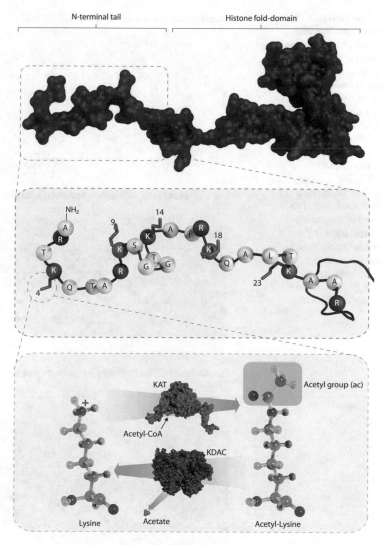

Fig. 3.2 Histone acetylation. Acetylation is shown as an example of a post-translational modification of histone proteins. On top, space-filling surface model with secondary structures of histone H3 is shown and in the center a zoom into its amino-terminal tail in the center. The basic amino acids lysine (K) and arginine (R) are indicated in blue. On the bottom, the activity of KATs removes the positive charge, while KDACs can reverse this process

nucleosomes. For example, H2A.Bbd lacks acidic amino acids at its carboxy terminus, as a consequence of which only 118–130 bp (versus 147 bp) of genomic DNA are wrapped around the respective histone octamer. This leads to the formation of less compact respectively more accessible chromatin and thus facilitates transcription.

Fig. 3.3 The impact of histone variants on chromatin. The incorporation of histone variants (*beige* and *brown*) into chromatin can result in homotypic or heterotypic nucleosomes (*top*) and can have direct (*center*) or indirect (*bottom*) effects on chromatin structure and function. Histone variants can directly influence nucleosome structure and stability, such as affecting the susceptibility for chromatin remodeling. The composition of the nucleosome can also have indirect effects on chromatin organization and function, for example, by inducing the recruitment of specific readers of histone modifications

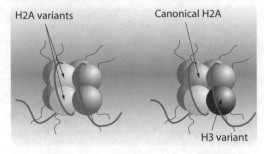

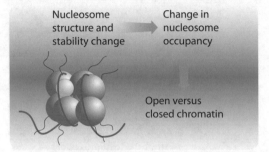

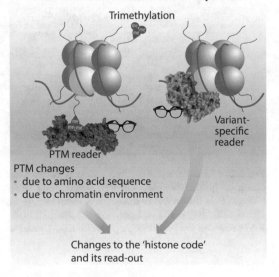

The same nucleosome may contain multiple histone variants in the same nucleosome. There are homotypic nucleosomes, which carry two copies of the same histone, and heterotypic nucleosomes, which contain a canonical histone and a variant histone or two different histone variants (Fig. 3.3). This allows for

Table 3.1 Main functions of major histone variants

Histone variant	% identity versus the canonical histone	Functions
H2A.X	96	DNA damage repair
H2A.Z	60-65	Transcriptional regulation, early embryonic development, chromosome stability
Macro H2A	64	Early embryonic development, enriched in the inactive X chromosome
CENP-A	highly divergent	Centromere-specific
H3.3	97	Early embryonic development, replaces H3 at transcriptional active regions

greater variability in nucleosome formation, stability and structure. For example, nucleosomes that contain H2A.Z and H3.3 are less stable than canonical nucleosomes and are often found at nucleosome-depleted regions of active promoters, enhancers and insulators. These labile H2A.Z/H3.3-containing nucleosomes serve as "place holders" and prevent the formation of stable nucleosomes around regulatory genomic regions. They can be easily displaced by transcription factors and other nuclear proteins that are not able to bind genomic DNA in the presence of a nucleosome composed of canonical histones. Thus, variable composition of nucleosomes can directly influence gene expression.

Chromatin modifiers can differentially recognize ("read") nucleosomes containing variant histones (Fig. 3.3). In this way variant histones extend the "histone code" (Sect. 5.1) to a "nucleosome code." For example, the protein microcephalin (MCPH1) specifically recognizes serine and tyrosine phosphorylation in H2A.X after DNA damage. In this way, MCPH1 serves as an early sensor of DNA damage and connects chromatin with the DNA repair machinery. Accordingly, distinct histone variant-interacting proteins and chromatin modifiers form a "histone variant network". The histone variants H2A.Z, macroH2A and H3.3 play a role in early embryonic development by regulating the lineage commitment of stem cells. Moreover, they are also involved in the process of cellular reprograming to pluripotency (Table 3.1, Sect. 8.3). Furthermore, changes in the deposition of histone variants affects the process of tumorigenesis, i.e. disturbances in the histone variant network can promote cancer development (Sect. 10.3). Taken together, histone variants play a relevant role in human health and disease and thus the components of the histone variant network can serve as targets for personalized diagnosis, prognosis and therapy.

3.4   Euchromatin and Heterochromatin

As already briefly outlined in Fig. 1.6, chromatin has various levels of organization that represent different layers of regulation. The first regulatory layer is the post-translational modification of histone tails (Fig. 3.4, *top left*), which is summarized as the "histone code". Figure 3.2 already presented a first example of the

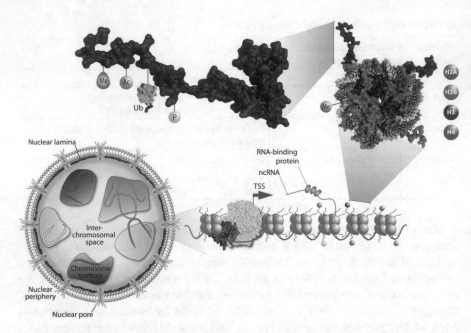

Fig. 3.4 Different layers of regulatory features in chromatin. The first regulatory layer of chromatin is the post-translational modifications of histone tails, such as methylation (Me), acetylation (Ac), ubiquitination (Ub) and phosphorylation (P) (see also Fig. 3.2) (*top left*). The second layer is represented by winding genomic DNA – which itself can be methylated on cytosine residues, around histone octamers – in order to form nucleosomes (*top right*). The third level is the positioning of nucleosomes in higher and lower density forming hetero- and euchromatin (*bottom right*). This influences the accessibility of genomic DNA for transcription factors, other nuclear proteins and ncRNAs, which in turn modulate the packaging status of the respective chromatin regions. Chromatin loops, such as TADs and LADs, and the positioning of chromosomes in spatial territories determines the nuclear architecture (*bottom left*). There are interactions within and between chromosomes, as well as between chromosomes and nuclear structures, such as nuclear pores, the inner nuclear membrane and the nuclear lamina

regulatory potential of these modifications and Chap. 5 will provide more details. The second layer of regulation is the formation of nucleosomes based on histone octamers (including histone variants) and genomic DNA (Fig. 3.4, *top right*). The modification of the latter at their cytosines is an additional central regulatory epigenetic feature and will be discussed in Chap. 4. The third regulatory layer is the density of chromatin packaging into hetero- and euchromatin influencing the accessibility of genomic DNA for transcription factors and other nuclear proteins and regulatory ncRNAs (Fig. 3.4, *bottom right*).

In the following, we will provide an overview on the impact of the transition of heterochromatin into euchromatin and vice versa. The architecture of the nucleus, which is based on chromatin loops, such as TADs and LADs, and the organization of chromosome into different territories of the nucleus, represents the fourth and largest-scale layer of chromatin regulation (Fig. 3.4, *bottom left*). This will be discussed in Sects. 3.5 and 3.6.

   Between cell divisions or when cells are terminally differentiated, they are in
the so-called "interphase." Although in this phase chromosomes are microscopi-
cally not visible, the nucleus is not a homogeneous organelle, but has defined
chromosomal regions, referred to as chromosome territories. For example, centromeres, telomeres and insulator assemblies, cluster with each other and other genomic regions and define distinct nuclear compartments. Moreover, there are nuclear
structures, such as the nuclear periphery, nuclear pores and the heterochromatic
compartment, and nuclear bodies, such as nucleoli, nuclear speckles and PML
bodies. Staining techniques show that in the interphase nuclei lighter areas, called
euchromatin, are found towards the center of the nucleus, and darker aggregates,
referred to as heterochromatin, in their periphery.

   In general, the more densely chromatin is packed, the less active it is.
Constitutive heterochromatin assembles on genomic DNA containing repetitive
sequences, such as centromers and telomers, where gene density is low. In contrast,
facultative heterochromatin is found on genes that are silenced in a given tissue or
cell type but need to keep their potential to be activated by appropriate signals, like
during cellular reprograming (Chap. 8). This implies that facultative heterochromatin
can reversibly transform into euchromatin, in order to facilitate gene expression.
Original euchromatin becomes condensed only during mitosis and is far more gene-
rich than heterochromatin.

   The protein HP1 is a key initiator in heterochromatin formation, since it recognizes histone marks of inactive, i.e. transcriptionally silent, regions of chromatin,
such as H3K9me3 or H3K27me3, and then recruits other heterochromatin components. HP1 also binds to other HP1 molecules attached to the chromatin fiber,
creating staple-like connections between HP1-associated chromatin domains, and
keeps the genomic region in a compact state. In this way, heterochromatin prevents unwanted transcription, stabilizes the genome and keeps cells in their (term-
inal) differentiation status. Despite its high level of compaction, heterochromatin
must remain accessible for enzymes, for example, in cases when DNA repair is
necessary. Moreover, in particular on regulatory genomic regions, such as enhancers and promoters, the formation of heterochromatin must be reversible (facultative heterochromatin). This implies that chromatin density plays an important role
in regulating gene expression, which involves a competition between transcription
factors and nucleosomes for critical genomic loci. Thus, chromatin is not static
but dynamic allowing and restricting transcription factor binding, i.e. it acts as a
"gatekeeper" of regulatory genomic regions.

   DNA accessibility not only differs within cell nuclei, but also between cell
types, and changes dynamically both during development and in response to external stimuli. Such changes in DNA accessibility are mediated by specialized transcription factors, such as the pioneer factor forkhead box (FOX) A1. These pioneer
factors have a DNA-binding sequence that is shorter than that of regular transcription factors. Therefore, FOXA1 and similar proteins are able to bind genomic DNA
even when it is covered by nucleosomes and make it accessible to other transcription factors. Alternatively, transcription factors might open DNA by competing
with nucleosomes for DNA binding. Furthermore, a large set of enzymes that either

post-translationally modify chromatin or remodel it by moving, reconfiguring or ejecting nucleosomes influence this competition (Chaps. 6 and 7).

Euchromatin is characterized by general acetylation of lysines in the tails of histones H3 and H4, H3K27ac and the tri-methylation of H3K4. In contrast, in heterochromatin H3K9, H3K27 and H4K20 are either mono-, di- or tri-methylated. The effects of chromatin modifiers, such as KATs and KDACs, are primarily local and may cover only a few nucleosomes up- and downstream of the starting point of their action. The same applies to chromatin remodeling enzymes, such as the SWI/SNF complex (Chap. 7) and lysine methyltransferases (KMTs, Chap. 6). Figure 3.2 has already indicated a scenario, in which the balance in the actions of KATs and KDACs decides, in which direction facultative heterochromatin turns.

1. In case there is more KAT activity, the chromatin is locally acetylated, the attraction between nucleosomes and genomic DNA decreases and the latter gets accessible for activating transcription factors, basal transcription factors and Pol II. In this euchromatin state chromatin remodeling enzymes, such as SWI/SNF, may have to fine-tune the position of the nucleosomes, in order to obtain full accessibility of the respective binding sites.
2. In the opposite case, when KDACs are more active, acetyl groups get removed and the packing of chromatin locally increases. KMTs methylate the same or adjacent amino acid residues in the histone tails that attract heterochromatin proteins, such as HP1, and further stabilize the local heterochromatin state.

3.5 Nuclear Architecture

The nuclear envelope is a membrane bilayer, in which nuclear pores are embedded. These pores have two roles: (i) they span the two membranes, and (ii) they mediate transport in both directions between the cytoplasm and the nucleus (Fig. 3.5). The inner side of the nuclear envelope is coated with nuclear lamina, which is a complex of lamins and a number of additional proteins. Lamins maintain the shape and mechanical properties of the nucleus. LAD-lamin interactions serve as a structural backbone for the organization of interphase chromosomes, i.e. they form a nucleoskeleton. LADs vary in size from 0.1–10 Mb and may cover up to 40% of the human genome. They are strongly enriched for H3K27me3 and H3K9me2 modifications, which are markers for Polycomb protein-repressed and heterochromatic genomic regions, respectively.

LADs have a low gene density, but in total they still contain thousands of genes, most of which are not expressed. Accordingly the nuclear periphery is enriched for heterochromatin, whereas euchromatin is found more likely in the center of the nucleus. This suggests that the location of a gene within the nucleus is a functionally important epigenetic parameter. Interestingly, during gene activation, sites of active chromatin, represented by H3K4me3, H3K36me3, H4K20me1 and H2NK5me1 modifications, are often in close spatial proximity and cluster at Pol II foci that are interpreted as transcription factories. These sites of open

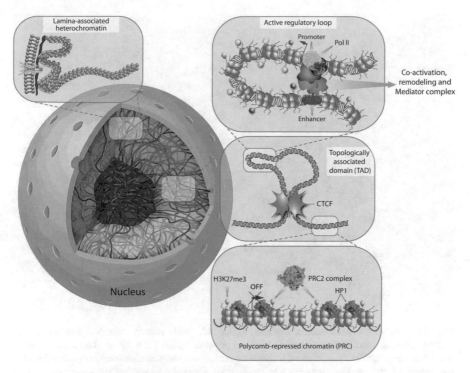

Fig. 3.5 Chromatin architecture. Mediated by structural proteins, chromatin within the nucleus forms a 3D architecture (*center left*). Heterochromatin is composed of stably repressed, inaccessible genomic elements and is located closer to the nuclear lamina (*top left*). Two CTCF proteins bound at adjacent chromatin boundaries form a complex with cohesin and other mediator proteins (*center right*). In this way, regulatory genomic regions, such as enhancers and promoters, that are separated by a genomic distance can get into physical contact within DNA loops (*top right*). TADs distinguish such genomic regions with active enhancers from chromatin tracts silenced by PRC1 and 2 (*bottom right*)

chromatin are found in the center of the nucleus (Fig. 3.6). However, in the process of cellular aging, referred to as senescence, which is represented by an irreversible exit from the cell cycle in response to exogenous and endogenous stress, heterochromatin re-localizes from the nuclear periphery to the interior (Sect. 9.3).

One major purpose of chromatin architecture is to bring regulatory genomic regions, such as enhancers and promoters, into close proximity. In this way, transcription factors, their co-factors and associated chromatin modifiers localized on enhancers get into contact with the mediator complex, basal transcription factors and Pol II bound to TSS regions (Fig. 3.5). CTCF can form chromatin loops between far distant binding sites, either by self-dimerization or via the recruitment of cohesin, a ring-shaped protein complex that can embrace two double-stranded DNA helices. Cohesin is well known for its central role in sister chromatid cohesion during S phase and maintaining it until mitosis. In a similar way cohesin stabilizes long-range contacts between enhancers and promoters. CTCF is also the

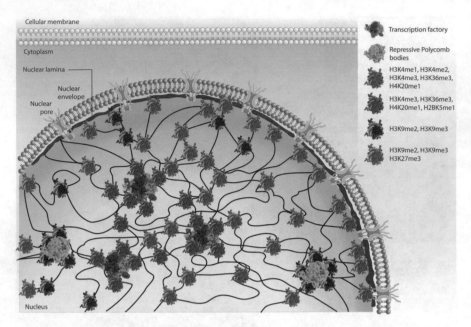

Fig. 3.6 Chromatin modification signatures associate with relative position features in the nucleus. Histone modifications correlate with the position within the nucleus: chromatin modifications that are generally associated with active transcription (*green nucleosomes*) are often found in the center of the nucleus, whereas chromatin with generally repressive modifications (*orange nucleosomes*) is associated with the nucleoskeleton. Regions with active modifications (*blue nucleosomes*) may participate in transcription factories (*purple Pol II in the center*). Blocks of histone H3K27me3 (*dark red nucleosomes*) may be components of Polycomb bodies (*yellow*)

main DNA-binding protein of insulator regions, which restrict the action of enhancers in activating distant genes. Interestingly, DNA binding of CTCF can be modulated by DNA methylation (Sect. 4.4).

The clustering of heterochromatin at the nuclear periphery creates silencing foci, so-called Polycomb bodies. These are complexes of members of the PcG family, such as the components of Polycomb repressive complex (PRC) 1 and 2, that act as transcriptional repressors being essential for maintaining tissue-specific gene expression programs, i.e. these proteins ensure the long-term controlled repression of specific target genes. H3K27me3 serves as a hallmark of facultative heterochromatin, since it is the major repressive histone modification (Fig. 3.6). In ES cells H3K27me3 marks are found at TSS regions of many key regulators of differentiation. Repression mediated by H3K27me3 involves both PRC1 and PRC2. H3K27me3 recruits PRC1 that catalyzes mono-ubiquitynation at H2AK119. This imposes a poised state of Pol II at repressed TSS regions (Chap. 6). Moreover, PRC1 mediates the compaction of chromatin marked by H3K27me3 that in turn recruits PRC2. In ES cells the loss of such repressive histone marks increases the chance of spontaneous differentiation. Long-range Polycomb protein-dependent chromatin interactions involve the *HOX* genes and extend across TADs and chromosomes.

3.6 TADs and Chromosome Territories

The position of chromatin, and with that the position of genes, is not fixed, but there are dynamic changes in the contacts between the nucleoskeleton and genomic DNA involving single genes or small gene clusters. These changes are most pronounced during development. For example, ES cells possess dispersed chromatin with limited compaction. During differentiation the cells show changes in their chromatin structure that include larger compaction of genomic domains, i.e. embryonic development proceeds from a single cell with dispersed chromatin to differentiated cells with nuclei that show compact chromatin domains being located in the periphery (Chap. 8). For example, the physical re-location of a gene from the nuclear periphery to the center unlocks it to be expressed in a future developmental stage. Moreover, also stress-induced genes show changes in their nuclear position. Upon activation, many genes move away from their original chromosome territory towards the inter-chromosomal space that is enriched for transcription factories. This fits also with the observation that gene-rich chromosome territories and active genes are found in the nuclear interior.

The two main experimental approaches for studying 3D genome organization are FISH and 3C-based methods (Sect. 2.1). FISH live-cell imaging provided dynamic views of chromatin domains, while 5C, Hi-C and ChIA-PET mapped the whole genome in kb resolution for chromatin loops, such as TADs and LADs. The latter methods are performed with large cell populations and thus provide a probabilistic view of chromosome folding and nuclear organization. In contrast, in FISH multiple individual cells are analyzed, i.e. both methods are complementary to each other and may be used for cross-wise validation. However, there is significant cell-to-cell and time-dependent variation in chromatin folding, so that the spatial distance between two genomic loci can show a rather wide distribution. 3C-based methods detect only events in cells, in which two loci are in close proximity, while FISH can determine the spatial distance between the loci in any cell. Thus, both methods investigate different subpopulations of cells and may lead to apparent inconsistencies. The 4D Nucleome project (https://commonfund.nih.gov/ 4Dnucleome/index) aims to overcome this problem via developing and employing a range of genomic, imaging and modeling methods to study 3D genome organization.

Global analyses of chromatin contacts in human cells, as performed by Hi-C, initially identified some 2,000 TADs with an average size of some 1 Mb (Sect. 7.3). Accordingly, most TADs contain a number of genes that may be regulated by the same set of enhancers within this genomic region, such as often observed for gene clusters. Individual TADs are separated by each other by boundary regions that are enriched both for enhancer markers, such as H3K4me1, and signs for repression, such as H3K9me3. Often TAD boundaries are identical with insulators and bind CTCF (Fig. 3.7), i.e. they separate functionally distinct regions of the genome from each other. In this way, CTCF is not only involved in smaller-scale DNA looping resulting in enhancer-promoter contacts (Figs. 3.5 and 3.7, *right*), but also in larger-scale loop

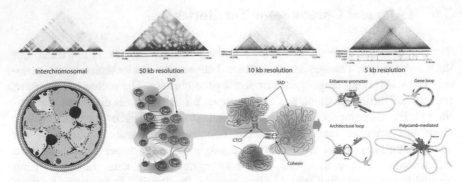

Fig. 3.7 Hierarchy of chromatin architecture. Hi-C data of four levels of resolution (*top*) are schematically interpreted (*bottom*) as inter-chromosomal interactions between intermingled chromosome territories (*left*), TADs with inter-domain interaction (*center left*), TADs (*center right*) and enhancer-promoter loops (*right*). Within Hi-C maps, chromatin loops and TADs are typically recognizable as contiguous triangles. These indicate that regions within the same TAD interact with each other more often than with regions of neighboring TADs. The distinction into TADs correlates with many features of the linear genome, such as patterns of histone modifications or gene expression as well as association with nuclear lamina

formation. Thus, TADs are the units of chromosomal organization and segregate the human genome into at least 2,000 domains containing co-regulated genes (Fig. 3.7, *center*). Based on higher resolution Hi-C maps the number of TADs may be five-times larger and their average size accordingly lower (Sect. 7.3).

Another level of chromatin architecture in the interphase nucleus is the location of whole chromosomes in separate chromosome territories, which seem to be separated by an inter-chromosomal compartment (Fig. 3.8). Chromosomes fold in their territories in a way that active and inactive TADs are found in distinct nuclear compartments. The active regions are preferentially located in the nuclear interior, whereas inactive TADs accumulate at the periphery. In addition, TADs that are heavily bound by tissue-specific transcription factors are in different neighborhoods than those interacting with PcG proteins (Fig. 3.6). Nevertheless, single-cell analysis (Box 2.1) suggests that on this level genome folding is highly probabilistic, i.e. a given TAD will have in each given cell different neighbors in space. Therefore, such higher order topological features of chromatin folding may only have a contributory, but not a deterministic role in gene regulation.

To some extent chromosome territories intermingle, which could explain inter-chromosomal interactions (Fig. 3.7, *left*). Nevertheless, interactions between loci on the same chromosome are much more frequent than contacts in *trans* between different chromosomes. Interestingly, the volumes of chromosome territories depend on the linear density of active genes on each chromosome, i.e. chromatin with higher transcriptional activity occupies larger volumes in the nucleus. Moreover, the overall level of chromosome intermingling correlates with the rate of transcription. This suggests that genome architecture and genes expression are closely linked.

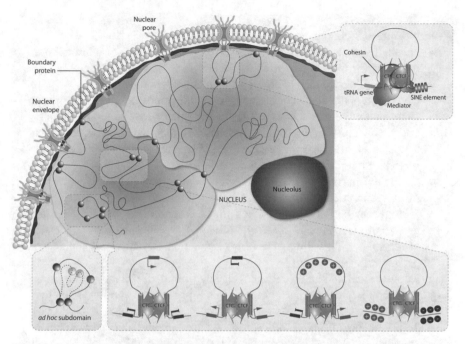

Fig. 3.8 Topological domains in the genome. In this scheme two chromosome territories (*blue and green*) are shown that each contain a chromosome. Different examples of smaller-scale chromatin loops are indicated, such as (i) CTCF-mediated enhancer-promoter contacts (*top right*), (ii) non-functional transient loops not involving CTCF (*bottom left*) and (iii) different CTCF-mediated constellations of insulating active (*green*) from inactive (*red*) chromatin regions (*bottom right*, plus and minus signs represent open and repressive chromatin, respectively). Interestingly, only some 15% of CTCF binding sites are found within a TAD boundary, i.e. the most of them are located within TADs and may be involved in intra-TAD interactions

Key Concepts
- Chromatin is the physical representative of the epigenome.
- Nucleosomes are the regularly repeating units of chromatin, but they can vary from one genomic region to the other by different post-translational modifications of histones and the introduction of histone variants.
- In contrast to transcription factors, the DNA binding of histones is not sequence specific.
- The tails and globular domains of histone proteins provide over 130 sites for post-translational modifications.
- The regular positioning of nucleosomes has the effect that the position of one nucleosome determines the location of its neighbors.
- Histone variants can directly influence nucleosome structure and stability, such as affecting the susceptibility for chromatin remodeling.
- The more densely chromatin is packed, the less active it is.

(continued)

Key Concepts (continued)

- Facultative heterochromatin can reversibly transform into euchromatin, in order to facilitate gene expression.
- H3K27me3 serves as a hallmark of facultative heterochromatin.
- Chromatin is not static but dynamic, allowing and restricting transcription factor binding, i.e. it acts as a gatekeeper of regulatory genomic regions.
- The architecture of the nucleus is based on chromatin loops, such as TADs and LADs, and the organization of chromosomes into different territories of the nucleus.
- One major purpose of chromatin architecture is to bring regulatory genomic regions, such as enhancers and promoters, into close proximity.
- The nuclear periphery is enriched for heterochromatin, whereas euchromatin is found more likely in the center of the nucleus, i.e. the location of a gene within the nucleus is a functionally important epigenetic parameter.
- Upon activation, many genes move away from their original chromosome territory towards the inter-chromosomal space that is enriched for transcription factories.
- TADs are the units of chromosomal organization and segregate the human genome into at least 2,000 domains containing co-regulated genes.
- Chromosomes are folded in their territories in a way that active and inactive TADs are found in distinct nuclear compartments.
- The volumes of chromosome territories depend on the linear density of active genes on each chromosome, i.e. chromatin with higher transcriptional activity occupies larger volumes in the nucleus.

Additional Reading

Bonev B, Cavalli G (2016) Organization and function of the 3D genome. Nat Rev Genet 17:661–678

Buschbeck M, Hake SB (2017) Variants of core histones and their roles in cell fate decisions, development and cancer. Nat Rev Mol Cell Biol 18:299–314

Carlberg C, Molnár F (2016) Mechanisms of Gene Regulation. Dordrecht: Springer Textbook. ISBN: 978-94-007-7904-4

Hnisz D, Day DS, Young RA (2016) Insulated neighborhoods: structural and functional units of mammalian gene control. Cell 167:1188–1200

Pombo A, Dillon N (2015) Three-dimensional genome architecture: players and mechanisms. Nat Rev Mol Cell Biol 16:245–257

Talbert PB, Henikoff S (2017) Histone variants on the move: substrates for chromatin dynamics. Nat Rev Mol Cell Biol 18:115–126

Chapter 4
DNA Methylation

Abstract The identity of each human cell is determined by its unique gene expression pattern. It must be remembered and passed on to daughter cells by epigenetic mechanisms, of which DNA methylation is the most prominent. Methylation of the fifth position of cytosine (5mC) in a CpG dinucleotide has a profound impact on genome stability, gene expression and development. DNA methylation is often associated with transcriptional silencing of repetitive DNA and genes that are not needed in a specific cell type. CpG islands are associated with the majority of human gene promoters. The oxidative modification of 5mC via the TET/thymine-DNA glycosylase (TDG) pathway provides a new perspective on how heritable DNA methylation patterns may be dynamically regulated. This is the mechanistic basis of translating DNA methylation into biological actions of DNA binding proteins that specifically recognize either unmethylated or methylated genomic DNA. Insulators are genomic loci that separate genes located in one chromatin region from promiscuous regulation by transcription factors binding to enhancers of neighboring chromatin regions. The methylation-sensitive genome regulator CTCF is the main protein binding to insulators. Moreover, CTCF-mediated loops at several developmentally regulated loci provide a mechanistic explanation of genomic imprinting. Furthermore, DNA methylation is highly informative when studying gene regulation in normal and diseased cells. For example, aberrant DNA methylation is a well-established marker of cancer leading to inactivation of tumor suppressor genes, disturbance in genomic imprinting and genomic instabilities through reduced heterochromatin formation on repetitive sequences.

In this chapter, we will present the impact of DNA methylation in the epigenomic processes, such as the formation of heterochromatin and subsequent gene silencing. We will present CTCF and other transcription factors as methylation-sensitive proteins. In this context will discuss different types of insulators and the role of DNA methylation in genomic imprinting.

Keywords DNA methylation · CpG islands · DNA methyltransferase · TET proteins · 5mC modifications · gene silencing · insulator · gene silencing · CTCF · imprinting · paternal gene · maternal gene

© Springer Nature Singapore Pte Ltd. 2018
C. Carlberg, F. Molnár, *Human Epigenomics*, https://doi.org/10.1007/978-981-10-7614-5_4

4.1 Cytosines and Their Methylation

Methylation of genomic DNA at position 5 of cytosines (5-methylcytosine (5mC))
is an intensively studied epigenetic modification, which in humans and other
mammals is primarily restricted to CpG dinucleotides (Box 4.1). Since only CpGs
can be symmetrically methylated, they are the exclusive methylation marks that
remain after DNA replication on both daughter strands. This property makes
CpGs most suited for long-term epigenetic memory that can even be inherited to
the next generation (Sect. 9.3). Less than 10% of all CpGs occur in genomic
regions with a CG density of more than 55% and referred to as CpG islands
(Fig. 4.1).

Box 4.1 Cytosines and their Methylation
The average CG base pair percentage of the human genome is 42%
(Fig. 4.1). In principle, each cytosine can be methylated at its 5th position,
but those of CG dinucleotides (CpGs) are functionally most important, in
particular, if they occur in clusters. CpG islands are defined as regions of at
least 200 bp showing a CG percentage of higher than 55%, but typically they
are 300–3,000 bp long. Nevertheless, also the methylation of CpH dinucleotides (H = A, C or T) seems to play a role in epigenetic memory (Chap. 11).

Most of CpGs remain methylated during development, but CpG islands located
close to the TSS regions of housekeeping or developmentally regulated genes
have a very low methylation status (Sects. 4.2 and 8.2). Interestingly, a C to T
transition at the location of CpG islands is one of the most frequent mutations
found in human diseases. This implies that DNA methylation reduces the efficiency of DNA repair resulting in the accumulation of mutations at these sites.

Cytosine methylation does not only occur at CpGs, but also at CpH dinucleotides. Non-CpG methylation (mCH) occurs in all human tissues, but is most common in long-lived cell types, such as stem cells and neurons. This type of
methylation can serve memory function and is negatively correlated with gene
activity (Sect. 11.1). Proteins that specifically bind methylated DNA ("readers"
Fig. 4.2), such as methyl-CpG-binding protein 2 (MeCP2), do not only interact
with methylated CpG sites but also with CpH loci.

DNMTs catalyze in a one-step reaction the cytosine methylation of genomic
DNA. DNMT1 is responsible for maintenance of DNA methylation during the replication process and works together with its partner ubiquitin-like plant homeodomain
and RING finger domain 1 (UHRF1), that preferentially recognizes hemi-methylated
CpGs. With the exception of imprinted genomic regions (Sect. 4.4) DNMT3A and
DNMT3B perform de novo (i.e. new) DNA methylation at early embryogenesis, i.e.
together with DNMT1 these enzymes are "writers" of DNA methylation (Fig. 4.2,
left and Fig. 4.3). In the absence of functional DNMT1/UHRF1 successive cycles of
DNA replication lead to passive loss of 5mC, such the global erasure of 5mC in the

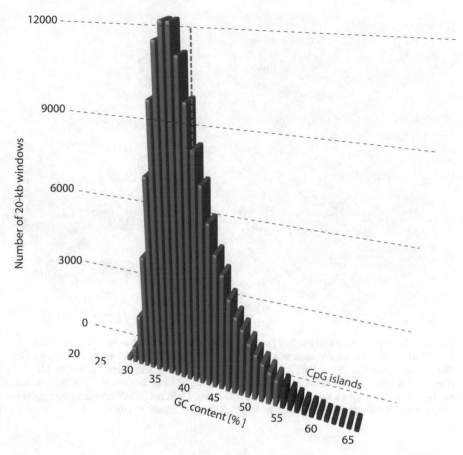

Fig. 4.1 CG content of the human genome. Due to subsequent passive deamination of genomic DNA, the average CG base pair content in the human genome is not 50% but only 42%. CpG islands (*red line*) have a CG percentage of 55%, i.e. only a minority of CpGs belong to CpG islands

maternal genome during pre-implantation (Sect. 8.1). Interestingly, the first approved epigenetic drug, decitabine (5-aza,2′-deoxy-cytidine), blocks DNA methylation via inhibiting DNMTs.

Active demethylation is a multi-step process that involves the methylcytosine dioxygenase enzymes TET1, TET2 and TET3 that convert 5mC to 5hmC (Fig. 4.2, *left* and Fig. 4.3). 5hmC is found in most cell-types but in levels of only 1–5% of that of 5mC. Adult neurons seem to be an exception, since their 5hmC level is 15–40% compared to 5mC rates (Sect. 11.1). In two further oxidation steps TETs convert 5hmC into 5-formylcytosine (5fC) and to 5-carboxylcytosine (5caC). 5fC and 5caC are significantly less prevalent (0.06–0.6% and 0.01% of 5mC, respectively) than 5hmC, i.e. TETs tend to halt at 5hmC. The oxidized cytosines are deaminated to 5-hydroxyuracil (5hmU) so that they create a 5hmU:G mismatch, which is

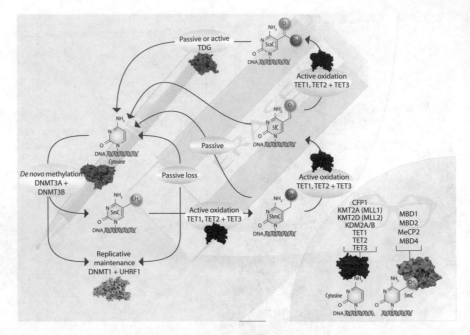

Fig. 4.2 Writing, erasing and reading cytosine methylations. DNMT1 as well as DNMT3A and DNMT3B catalyze the methylation of cytosines at position 5, i.e. they act as "writers" (*left*). The dioxygenase enzymes TET1, TET2 and TET3 oxidize 5mC to 5hmC and further to 5fC and 5caC, which leads via the action of the DNA glycosylase TDG to the loss of DNA methylation, i.e. both types of enzymes function as "erasers" (*center*). Different sets of proteins either specifically recognize unmethylated cytosines or 5mC, i.e. they are "readers" (*right*)

recognized and removed by the glycosylase TDG (Fig. 4.2, *top* and Fig. 4.3). The abasic site is then repaired by the base excision repair (BER) machinery, which results in the overall demethylation of a specific cytosine.

The DNA methylation pattern of somatic cells represents an epigenetic program maintaining global repression of the genome and specific settings of imprinted genes. Thus, it is important to maintain the DNA methylome during replication via the action of DNMT1 and UHRF1. However, DNA demethylation occurs during specific developmental stages, such as the pre-implantation phase (Sect. 8.1) and development of primordial germ cells (PGCs, Sect. 8.3). This sets up pluripotent states in early embryos and erases parental-origin-specific imprints in developing PGCs.

4.2 DNA Methylation at the Genome Scale

Genome-wide maps of 5mC and its oxidized modifications, i.e. studies of the DNA methylome, provide an unbiased view of their roles in epigenome function and transcriptional regulation. Base-resolution mapping methods, such as bisulfite

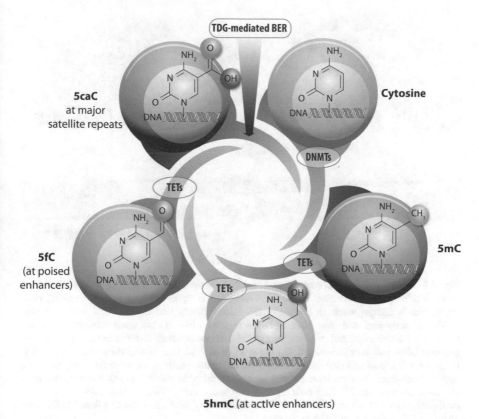

5hmC (at active enhancers)

Fig. 4.3 Methylcytosine derivatives. DNMTs are the enzymes responsible for cytosine methylation. However, DNA demethylation is a multi-step process that involves TET proteins that convert 5mC to 5hmC. These dioxygenase enzymes can further oxidize 5hmC to 5fC and to 5caC. TDG-mediated BER of 5fC and 5caC can regenerate unmethylated cytosines

sequencing, and affinity-based methods, such as MBD-seq and MeDIP-seq, were already outlined in Sect. 2.1. Both methods consistently identify genomic regions, in which cytosines are frequently modified, but they show discrepancies for sparsely distributed or infrequently modified cytosines, respectively. Anyway, the profiling of DNA methylomes has been extensively carried out under various physiological conditions in many different tissues and cell types. The majority (70–80%) of all CpGs are methylated across all tissues and cell types, with unmethylated CpGs mainly found in CpG islands close to TSS regions. During cellular differentiation, the methylation status of some 20% of all CpGs is dynamically modified. Thus occurs mainly at distal enhancers and contributes to tissue-specific gene expression programs. These changes involve cross-talk with a number of histone modifications, such as H3K9me3 and H3K4me3 (Chap. 5). Thus, the DNA methylome is an essential component of the epigenome as defined by IHEC (Sect. 2.2), and many datasets are already publically available.

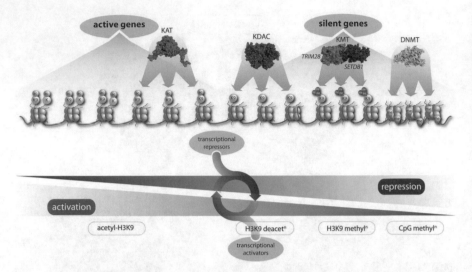

Fig. 4.4 Active and silent chromatin stages. DNA methylation and histone modification have different roles in gene silencing. While DNA methylation represents a very stable silencing mark (*right*) that is seldom reversed, histone modifications mostly lead to labile and reversible transcriptional activation and repression (*left*). At regions where histones are acetylated, genomic DNA remains unmethylated, whereas in repressed regions, where histones are methylated, also genomic DNA becomes methylated. On the level of both histone modification and DNA methylation there is a gradual transition between these extremes, i.e. there are more different epigenetic stages of chromatin than euchromatin and heterochromatin. Moreover, this also implies that there is a close co-operation between both types of chromatin marks: Histone methylation is involved in directing DNA methylation patterns, while DNA methylation can serve as a template for some histone modifications after DNA replication

During development, patterns of gene repression are established by both DNA methylation and post-translational modification of histone proteins (Chap. 5). While histone methylation causes easily reversible local formation of heterochromatin, DNA methylation mostly leads to stable long-term repression (Fig. 4.4). The human genome contains some 28 million CpGs, the majority of which are methylated and are located within regions of constitutive heterochromatin on endogenous transposable elements, such as short and long interspersed elements (SINEs and LINEs) and long-terminal repeats (LTRs), representing some 40% of the human genome (Box 4.2). LINEs and LTRs carry strong promoters that must be constitutively silenced, in order to prevent their activity. Therefore, these genomic regions are generally hyper-methylated. This silencing process is initiated by H3K9 methylation mediated by the KMT SETDB1, which in turn is recruited by tripartite-motif-containing protein (TRIM) 28. Then DNMTs are recruited, and the resulting DNA methylation operates as a long-term stabilizer of the transcriptional repression of the respective genomic regions. The silencing of transposable elements happens primarily during early embryogenesis, while in adult tissues de novo silencing is mediated by MeCP2 together with KDACs.

Box 4.2 Repetitive DNA in the Human Genome
About 50% of the sequence of the human genome is repetitive DNA, sorted
into the following categories (by order of frequency):

LINEs (500–8,000 bp)	21%
SINEs (100–300 bp)	11%
Retrotransposons, such as LTRs (200–5,000 bp)	8%
DNA transposons (200–2,000 bp)	3%
Minisatellite, microsatellite or major satellite (2–100 bp)	3%

The interspersed elements LINEs and SINEs are identical or nearly identical
DNA sequences that are separated by large numbers of nucleotides, i.e. the
repeats are spread throughout the whole human genome. LTRs are characterized by sequences that are found at each end of retrotransposons. DNA
transposons are full-length autonomous elements that encode for a transposase, by which the sequence can be moved from one position to another.
Microsatellites are often associated with centromeric or pericentromeric
regions and are formed by tandem repeats of 2–10 bp in length.
Minisatellites and major satellites are longer, with a length of 10–60 bp or
up to 100 bp, respectively.

The DNA methylation pattern of the human genome is bimodal (Fig. 4.5):
CpG-rich TSS regions ("high CpG-content promoters") and imprint control
regions (ICRs) have a low methylation level, while the default state of the remaining CpGs is methylated. In general, there is an inverse correlation between DNA
methylation of regulatory genomic regions and gene expression, i.e. in most cases
genome-wide analysis confirmed that methylated DNA is transcriptionally
repressed. However, analyses of changes in gene expression profiles in relation to
DNA methylation showed that a substantial portion of methylated CpGs are positively correlated with gene expression (Fig. 4.6, *left*), i.e. highly expressed genes
show high levels of DNA methylation in their gene bodies. Genes driven by high
CpG-content promoters are silenced when methylated, while genes with no CpG
islands close to their TSS regions ("low CpG-content promoters") are most likely
not regulated by DNA methylation at all. The differential methylation of the
human genome is established by de novo methylation combined with active
demethylation of CpG islands. During early embryogenesis, i.e. in the pre-
implantation phase, most CpGs are unmethylated. After implantation DNMT3A
and DNMT3B in combination with DNMT3L de novo methylate those CpGs that
had not been packed with H3K4me3-marked nucleosomes (Fig. 4.5, *top left*). In
contrast, H3K4me3-marked CpGs on TSS regions of CpG-content promoters stay
unmethylated (Fig. 4.5, *top right*).

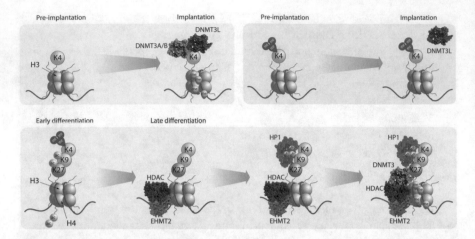

Fig. 4.5 Bimodal methylation of the human genome. In the pre-implantation phase, most CpGs in the embryonic genome are unmethylated. After implantation CpGs that are covered with unmarked nucleosome are specifically recognized by DNMT3L and then get de novo methylated by DNMT3A and DNMT3B (*top left*). In contrast, CpGs on H3K4me3-marked nucleosomes are not recognized by DNMT3L and accordingly stay unmethylated (*top right*). Later in differentiation, genes for key transcription factors determining cell lineage ("pluripotency genes") need to be inactivated (*bottom*). The KMT EHMT2 leads to H3K9 methylation at unmethylated CpGs on TSS regions of these genes, the attraction of HP1, recruitment of DNMT3A and DNMT3B, de novo DNA methylation and finally transcriptional silencing

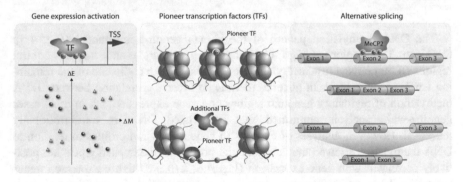

Fig. 4.6 Regulating biological processes by DNA methylation. Many transcription factors bind to methylated DNA (*left*). Genome-wide profiling of changes in gene expression (ΔE) and methylation levels (ΔM) demonstrated that the methylation of CpGs either positively (*blue triangles*) or negatively (*red dots*) correlate with gene expression. Transcription factors that bind methylated genomic DNA when wrapped around nucleosomes have a high likelihood being pioneer transcription factors (*center*). Methylation-sensitive transcription factors binding to exons, such as MeCP2, can regulate alternative splicing (*right*). When the exon is not methylated, MeCP2 does not bind and the exon is excluded

Pluripotency genes encode for key transcription factors that determine the cell lineage, such as *POU5F1* (also called *OCT4*) and *NANOG* (Sect. 8.3). In later stages of embryogenesis some of these genes need to be inactivated (Fig. 4.5, *bottom*). At the active stage during early embryogenesis, unmethylated CpGs at TSS regions of these genes are acetylated at histones H3 and H4 and methylated at H3K4. During differentiation, the KMT EHMT2 introduces H3K9 methylation attracting HP1, which leads to the recruitment of DNMT3A and DNMT3B and results in silencing de novo DNA methylation. In cases, when in differentiated cells pluripotency genes are silenced only by histone modification, these cells can be rather easily converted to iPS cells (Sect. 8.3). In contrast, after CpGs at the TSS regions of these genes are methylated, cellular reprograming is nearly impossible without altering key factors in the cell. Thus, there are different forms of gene silencing ranging from flexible repressor-based mechanisms to a highly stable inactive state being maintained by DNA methylation.

A parallel mechanism for inactivating CpGs involves promoter regions that are marked by H3K27me3 via PcG proteins for de novo methylation by DNMT3A and DNMT3B in combination with DNMT3L. This means that DNA methylation is interconnected with Polycomb protein-mediated silencing that can be maintained through cell division. Nevertheless, genome-wide studies demonstrate that during embryogenesis the CpGs of most genes remain largely unmethylated, while genomic region in some 2 kb distance from the CpGs often get methylated. These so-called CpG island shores are evolutionary conserved, and their methylation pattern correlates in a tissue-specific fashion with gene silencing.

DNA methylation is also a control mechanism for the activity of ICRs for mono-allelic expression of more than 100 maternally and paternally controlled genes. For example, there are different rates of de novo DNA methylation at imprinted maternal and paternal loci, suggesting that epigenetic memory reinstructs methylation. Furthermore, DNA methylation is involved in gene imprinting on the level of whole chromosomes, such as X chromosome inactivation in female cells (Sect. 4.4).

In principle, DNMT and TET enzymes can modify each of the approximately 28 million CpGs in the human genome to one of the five epigenetic states (C/5mC/5hmC/5fC/5caC, Fig. 4.3). Accordingly, the number of possible different DNA methylomes is immense and provides a large potential for information storage in the epigenome (Sect. 11.1). The general methylation level of CpG islands is 80–90%, while that of 5hmC modifications is only 1–30% and of 5fC and 5caC as low as 8–10%. Moreover, the different types of modifications are enriched at certain genomic regions, such 5hmC at active enhancers and intra-genic regions, 5fC at poised enhancers and 5caC at major satellite repeats (Box 4.2). 5fC and 5caC are both enriched at actively transcribed gene bodies, where they may regulate the rate of Pol II transcription. Thus, 5mC modifications are not only intermediates of the biochemical pathway restoring unmethylated DNA but seem to act themselves as epigenetic marks and the involved enzymes are epigenetic modulators. For example, TET proteins and oxidized 5mC establish and maintain transcriptionally poised or inactive states at lineage-specifically expressed genes, such as maintaining partial methylation states by mediating active demethylation, recruiting co-repressor complexes, such as SIN3A (SIN3 transcription regulator family member A), and modulating the binding of methylation-sensitive transcription factors (see Sect. 4.3).

4.3   Interaction of Transcription Factors with Methylated DNA

The basis for a mechanistic understanding of the DNA methylome is the specific interaction of transcription factors and other DNA-binding proteins with methylated and unmethylated genomic DNA. These "readers" (Fig. 4.2) translate the methylation signal into activities of biological processes. For example, in the process of alternative splicing, MeCP2 binding to methylated genomic DNA of an exon leads to its inclusion into mRNA, while in the absence of MeCP2 binding the exon is excluded (Fig. 4.6, *right*). The central point in this translation process is that methylation and demethylation of CpGs modulate the DNA-binding affinity of transcription factors, i.e. DNA methylation is a signal that is differentially recognized by specific protein domains. A combined epigenomics-proteomics approach demonstrated that nearly half of all 1,600 human transcription factors have a DNA methylation preference. 34% of these transcription factors are positively affected by methylation, such as several homeodomain transcription factors, half of which do not bind DNA when is unmethylated, and only 23% of them are negatively influenced.

The MBD proteins MeCP2, MBD1, MBD3 and MBD4 are the master examples of proteins binding symmetrically methylated CpGs, but they have no sequence specificity. This means that MBD proteins are not classical transcription factors, but act as adaptors for the recruitment of chromatin modifiers, such as KDACs and KMTs, to methylated genomic DNA. However, also without a MBD many transcription factors recognize methylated DNA, such as ZBTB33 (also known as Kaiso), that binds a specific methylated sequence with its C2H2 zinc-finger domain. Moreover, the pioneer transcription factor CEBP α, the zinc-finger protein (ZFP) 57 and its co-factor TRIM28 interact with specific methylated sequences. Some transcription factors, such as OCT4, SRY-box 2 (SOX2) and Kruüppel-like factor 4 (KLF4), specifically bind methylated DNA even when it is wrapped around nucleosomes (Fig. 4.6, *center*). They act as pioneer transcription factors by recruiting chromatin modifiers that help to open the heterochromatic DNA, change its status to euchromatin and initiate transcription. Like 5mC, also 5hmC is a signal that influences transcriptional activity by attracting or repelling specific transcription factors. For example, MBD3, MBD4 and MeCP2 bind to hemi- or fully 5hmC-marked CpGs, i.e. these proteins have a dual binding capacity for both 5mC and 5hmC.

There are a number of chromatin-associated proteins that have a strong preference for unmethylated CpGs. Some of them, such as TET1 and TET3, carry a zinc-finger cysteine-X-X-cysteine (CXXC) domain. Other examples are CXXC finger protein 1 (CFP1) and the histone demethylases (KDMs) KDM2A and KDM2B (Fig. 4.2, *right*). CPF1 binds unmethylated CpGs and recruits KMTs that leave H3K4me3 marks at active promoters. DNMT3A and DNMT3B recognize unmethylated CpGs but are allosterically inhibited by H3K4 methylation (Fig. 4.5, *top right*). In this way, genomic DNA at these regions is kept unmethylated and the chromatin active. In parallel, the histone variant H2A.Z is strongly enriched at unmethylated, active promoters. The H3K4-KMTs KMT2A and KMT2D (also known as mixed-lineage leukemia (MLL) 1 and 2) also carry CXXC domains, bind to unmethylated TSS of developmental genes and protect them from DNA methylation. This is another example how the interplay of histone modification and histone variants keeps the DNA methylation status low at selected genomic regions.

A very special methylation-sensitive transcription factor is CTCF, which has an essential role in imprinting control (Sect. 4.4). When CTCF binds to unmethylated ICRs in maternal alleles, distal enhancers cannot activate downstream genes. In contrast, when the ICRs in paternal alleles are methylated, CTCF does not bind downstream genes are activated via distal enhancers.

Classical transcription factor have no enzymatic activity that may methylate or demethylate CpGs, but they provide sequence-specific guidance to recruit DNMTs or TET proteins. This recruitment may often function indirectly via histone modification. A typical example is that of the *FOXP3* gene regulation during regulatory T cell proliferation (Fig. 4.7). In this case, the transcription factors CBFB (core-binding factor subunit-β) and RUNX1 (runt-related transcription factor 1) both bind a proximal enhancer downstream of the *FOXP3* TSS and stimulate prominent *FOXP3* mRNA expression. The binding of the transcription factors to the enhancer leads to its rapid demethylation allowing *FOXP3* protein binding for stable auto-regulation. Thus, transcription factor-mediated local demethylation of an enhancer region creates epigenomic memory that stabilizes T cell lineage progression over multiple cell divisions.

4.4 Insulators, CTCF and Imprinting

Insulators are genomic regions that block enhancer activities, i.e. they restrict the communication between enhancer and TSS regions. In addition to the prevention of overboarding enhancer activity, insulators can act as boundary elements that inhibit spreading of heterochromatin from silenced genomic regions to transcriptionally active parts of the genome. This means that boundary elements "insulate" closed from open chromatin, and inactive from active genes, respectively. Mechanistically insulators work by either forming a DNA loop or just serving as a neutral boundary to neighboring regulatory elements. Insulators are bound by the transcription factor CTCF, which is not only implicated in blocking of enhancer activity and heterochromatin spreading but also in 3D organization of the genome (Sect. 3.5).

CTCF carries the unusually high number of eleven zinc-fingers in its DNA-binding domain. The combinatorial use of these zinc-fingers creates a conformation that allows CTCF to recognize a large variety of DNA sequences and provides the transcription factor with a versatile role in genome regulation. CTCF

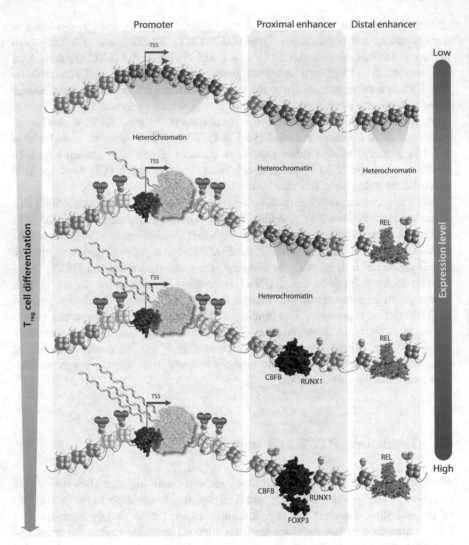

Fig. 4.7 Epigenetic memory of transcriptional activity. During differentiation of regulatory T cells, the *FOXP3* gene must show stable and strong expression. A homodimer of the transcription factor REL binds to a downstream enhancer and stimulates *FOXP3* mRNA expression and demethylation of the promoter region. *FOXP3* gene expression is stabilized through the binding of the transcription factors CBFB and RUNX1 to a more proximal enhancer. The latter induces local demethylation and permits binding of *FOXP3*. Thus, in an auto-regulatory fashion *FOXP3* ensures constitutive activity of the promoter of its own gene. The demethylation of its enhancer and promoter regions serves as an epigenetic memory of transcriptional activity through mitosis

is ubiquitously expressed in basically all human tissues. The genome-wide profiling of CTCF indicates in average some 30,000 binding sites, approximately 15% of which are involved in anchor sites for TADs (Sect. 3.6). The binding pattern of CTCF shows a higher level of conservation between tissues and cell types than most other transcription factors indicating that there is evolutionary pressure on the genomic structures formed with the help of CTCF. Like TSS regions, CTCF sites are often nucleosome-depleted, marked by specific histone modifications and kept in this state through the surveillance of chromatin modifying and remodeling enzymes. This suggests that insulators with CTCF binding sites are evolutionary derived from promoters and both types of genomic regions still use related mechanisms, in order to mediate their distinct functions. For example, a central function of TSS regions is their interaction with distal enhancer regions via DNA looping. Similarly, also for insulators their long-range communication between each other on one side and with enhancer and TSS regions on the other side is a key mechanism of their function.

Genomic CTCF binding is altered by methylation of CpGs within and around its core consensus sequence, i.e. CTCF binding to methylated sites is drastically reduced. Since higher-order chromatin structures stabilized by CTCF binding represent a form of epigenetic memory, this information storage can be modulated by DNA methylation. CTCF retains its information content by staying bound to DNA even during disruptions in chromatin structure being caused by DNA replication, transcription and chromatin compaction during mitosis. However, only a small subset of unmethylated CTCF binding sites keep CTCF proteins bound throughout the cell cycle, in order to protect these sites against de novo methylation. Thus, only those higher-order chromatin structures that are mediated by unmethylated CTCF sites can be inherited through mitosis. In contrast, when critical CTCF sites are methylated, for example, in response to developmental or environmental signals, the resulting abrogated CTCF binding will prevent the inheritance of the respective 3D chromatin structure. This means that CTCF-mediated chromatin structures represent a heritable component of phenotype-specific transcriptional and epigenetic programs.

In diploid cells the process of imprinting regulates for at least 100 genes of the human genome, which are expressed either in a paternal or in a maternal-specific manner   (www.geneimprint.com/site/genes-by-species.Homo+sapiens.imprinted-All). This is important for normal mammalian growth and development. Most imprinted genes were found as clusters, a master example of which is the chromosome 11p15 region that contains the protein-coding genes insulin-like growth factor 2 (*IGF2*), potassium voltage-gated channel subfamily Q member 1 (*KCNQ1*) and cyclin-dependent kinase inhibitor 1C (*CDKN1C*), as well as the ncRNA genes *H19* and *KCNQ1OT1* (Fig. 4.8, *top*). This genomic locus contains two ICR sites and is regulated by enhancers downstream of the *H19* gene. In maternally controlled alleles, ICR1 is unmethylated and binds CTCF, while ICR2 is methylated and not bound (Fig. 4.8, *center*). During post-implantation development CTCF binding is essential, in order to maintain the hypo-methylated state and to protect from de novo methylation in oocytes. CTCF blocks the long-range communication of an enhancer with the TSS region of the *IGF2* gene but allows the initiation of *H19* transcription. This results in the expression of *H19*, *KCNQ1* and *CDKN1C*, as well as in the repression of *IGF2* and *KCNQOT1*. In contrast, in paternally controlled alleles ICR1 is methylated and does not bind CTCF, while ICR2 is

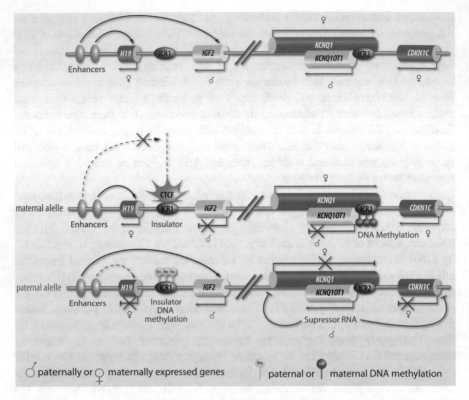

Fig. 4.8 Control mechanisms of the 11p15 imprinted cluster. The general structure of the 11p15 cluster (*top*) as well as the scenarios of maternally (*center*) and paternally (*bottom*) controlled alleles are illustrated. The Silver-Russell syndrome and the Beckwith-Wiedemann syndrome both are imprinting disorders that relate to this locus. *IGF2* encodes for a growth factor, *H19* for a long ncRNA limiting body weight, *KCNQ1* for a potassium channel, *KCNQ1OT1* for an antisense transcript of *KCNQ1* that interact with various chromatin components and *CDKN1C* for a cell cycle inhibitor. More details are provided in the text

unmethylated (Fig. 4.8, *bottom*). This reverses the expression pattern, so that *IGF2* and *KCNQOT1* are transcribed and *H19*, *KCNQ1* and *CDKN1C* not. The physiological consequence of this imprinting is that in maternally controlled cells growth and cell cycle are limited, while paternally controlled cells are primed for maximal growth.

The sensitivity of CTCF binding sites for DNA methylation provides an important mechanism for genomic imprinting. Moreover, the protein maintains allele-specific imprints and 3D organization of the locus during mitosis. Another well-studied example of imprinting is the inactivation of one X chromosome in female cells.

The inactive X chromosome is observed as Barr body in female interphase cells. The epigenetic process behind X chromosome inactivation is the long ncRNA *Xist* (X-inactive specific transcript), which is exclusively expressed from the X inactivation center of the inactive X chromosome. This represents a special form of imprinting that affects a whole chromosome. At the blastocyte stage, i.e. an early embryonic stage of approximately 100 cells (Sect. 8.1), one of the two X chromosomes is randomly inactivated in each cell. *Xist* RNA directs chromatin and transcriptional change by binding of the PRC2 complex, which leads to H3K27me3 heterochromatin marks throughout the whole X chromosome. This restricts the accessibility of DNA for transcription factors and their co-regulators being sufficient to silence all genes on the chromosome. At a later post-implantation stage many of these sequences undergo de novo methylation, but this happens after the genes of the X chromosome are already silenced.

4.5   Link of DNA Methylation to Disease

The Silver-Russell syndrome, a disease leading to undergrowth and asymmetry, and the Beckwith-Wiedemann syndrome, a disease leading to overgrowth, are imprinting disorders that relate to epigenetic errors in 11p15 (Fig. 4.8). For example, most of the patients with Beckwith-Wiedemann syndrome lost the methylation at ICR2, resulting in bi-allelic expression of the *KCNQ1OT1* ncRNA and aberrant repression of *CDKN1C*. Other Beckwith-Wiedemann syndrome patients show overexpression of *IGF2* caused by deletions in ICR1 on the maternal allele, disrupting CTCF binding and leading to bi-allelic *IGF2* and loss of *H19* expression. Many individuals with Silver-Russell syndrome have an opposite epigenetic phenotype, where ICR1 is unmethylated, resulting in bi-allelic *H19* expression and loss of *IGF2* expression.

Imprinting disorders are very rare, while perturbations in the DNA methylome are often observed in cancer (Sect. 10.2). For example, the CpG-rich promoters of tumor suppressor genes, such as *TP53*, are frequently hyper-methylated leading to gene silencing. Moreover, in acute myeloid leukemia (AML) the *DNMT3A* gene is very frequently mutated (in about 25% of adults with the disease). A loss of function of DNMT3A disturbs the methylome and makes the regulatory landscape of pre-leukemic blood stem cells more vulnerable for additional mutations. Similarly, mutations in the genes of chromatin modifiers may enhance the phenotype of cancer-driven mutations in transcription factor genes or their genomic binding sites.

DNA methylation profiles, for example, of white blood cells, may serve as biomarkers for evaluating the risk of cancer and a number of other diseases, such as the metabolic syndrome (Chap. 13). For example, mutations in the *MECP2* gene cause the neurodevelopmental disorder Rett syndrome, which belongs to the autism spectrum disorders (Sect. 11.2). Although biomarkers often do not explain the causality of a disease, they can monitor the disease state and may suggest appropriate therapy. Thus epigenomic profiles, such as DNA methylation patterns, in combination with genetic predisposition and environmental exposure, may be prognostic for personal risk of disease onset.

Key Concepts

- DNA methylation is the most prominent epigenetic mechanism to memorize cell identity and pass this information to daughter cells.
- CpGs can be symmetrically methylated and therefore are the exclusive methylation marks that remain after DNA replication on both daughter strands.
- CpGs that occur in genomic regions with a CG density of more than 55% are referred to as CpG islands.
- The majority (70–80%) of all CpGs are methylated across all tissues and cell types, with unmethylated CpGs mainly found in CpG islands close to TSS regions.
- Non-CpG methylation (mCH) occurs in all human tissues, but is most common in long-lived cell types, such as stem cells and neurons.
- DNMT1 is responsible for maintenance of DNA methylation during the replication process, while DNMT3A and DNMT3B perform de novo DNA methylation.
- The dioxygenase enzymes TET1, TET2 and TET3 oxidize 5mC to 5hmC and further to 5fC and 5caC, which leads via the action of the DNA glycosylase TDG to the loss of DNA methylation.
- 5mC modifications are not only intermediates of the biochemical pathway restoring unmethylated DNA, but act themselves as epigenetic marks and the involved enzymes are epigenetic modulators.
- There are different forms of gene silencing ranging from flexible repressor-based mechanisms to a highly stable inactive state being maintained by DNA methylation.
- DNA methylation is a control mechanism for the activity of ICRs for mono-allelic expression of more than 100 maternally and paternally controlled genes.
- The epigenetic process behind X chromosome inactivation is the long ncRNA *Xist*, which directs chromatin and transcriptional change.
- Methylation and demethylation of CpGs modulate the DNA-binding affinity of transcription factors, i.e. DNA methylation is a signal that is differentially recognized by specific protein domains.
- Insulators are genomic loci that separate genes located in one chromatin region from promiscuous regulation by transcription factors binding to enhancers of neighboring chromatin regions.
- The methylation-sensitive genome regulator CTCF is the main protein which binds to insulators.
- CTCF-mediated loops at several developmentally regulated loci provide a mechanistic explanation of genomic imprinting.
- CTCF-mediated chromatin structures represent a heritable component of phenotype-specific transcriptional and epigenetic programs.

(continued)

Key Concepts (continued)
- Aberrant DNA methylation is a well-established marker of cancer leading to inactivation of tumor suppressor genes, disturbance in genomic imprinting and genomic instabilities through reduced heterochromatin formation on repetitive sequences.
- DNA methylation profiles, for example, of white blood cells, may serve as biomarkers for evaluating the risk of cancer and a number of other diseases, such as the metabolic syndrome.

Additional Reading

Carlberg C, Molnár F (2016) Mechanisms of Gene Regulation. Dordrecht: Springer Textbook. ISBN: 978-94-007-7904-4

Du J, Johnson LM, Jacobsen SE et al (2015) DNA methylation pathways and their crosstalk with histone methylation. Nat Rev Mol Cell Biol 16:519–532

Edwards JR, Yarychkivska O, Boulard M et al (2017) DNA methylation and DNA methyltransferases. Epigenetics Chromatin 10:23

Schübeler D (2015) Function and information content of DNA methylation. Nature 517:321–326

Wu H, Zhang Y (2014) Reversing DNA methylation: mechanisms, genomics, and biological functions. Cell 156:45–68

Wu X, Zhang Y (2017) TET-mediated active DNA demethylation: mechanism, function and beyond. Nat Rev Genet 18:517–534

Zhu H, Wang G, Qian J (2016) Transcription factors as readers and effectors of DNA methylation. Nat Rev Genet 17:551–565

Chapter 5
The Histone Code

Abstract Post-translational modifications of histones are frequent and important epigenetic signals that control many biological processes, such as cellular differentiation in the context of embryogenesis. Acetylations and methylations of lysines in histone tails are understood best and may be most important epigenetic marks, but there are also a number of other acylations, such as formylation, propionylation, malonylation, crotonylation, butyrylation, succinylation, glutarylation and myristoylation, the functional impact of which is far less understood. In addition, there are phosphorylations at tyrosines, serines, histidines and threonines, ADP ribosylations at lysines and glutamates, citrullinations of arginines, hydroxylations of tyrosines, glycations of serines and threonines as well as sumoylations and ubiquitinations of lysines. All together these modifications form a collection of signals, referred to as the histone code, that mark genomic regions and pass information to chromatin modifiers and other nuclear proteins. The collection of histone modification patterns from various human cell lines and primary tissues, as provided by ENCODE, Roadmap Epigenomics and IHEC, provide a genome-wide basis of the histone code and its impact. Furthermore, signals from histone modifications combine with information provided by DNA methylations and post-translational modifications of non-histone proteins, such as transcription factors.

In this chapter, we will discuss the respective chemical and structural basis of different types of histone modifications and will present results from their genome-wide profiling. Finally, we will analyze the combination of histone modifications with DNA methylation and post-translational modifications of non-histone proteins.

Keywords Histone code · post-translational histone modification · histone methylation · histone acetylation · genome-wide profiling · DNA methylation · non-histone proteins

© Springer Nature Singapore Pte Ltd. 2018
C. Carlberg, F. Molnár, *Human Epigenomics*, https://doi.org/10.1007/978-981-10-7614-5_5

5.1   Molecular Principles of the Histone Code

Reversible post-translational modifications, such as phosphorylation, acetylation, and methylation, of key amino acid residues seem to be the major mechanisms of communication and information storage of proteins in the control of signaling networks in cells. A master example of such information-processing circuits is provided by the large set of modifications of the nucleosome in forming the histone proteins H2A, H2B, H3 and H4 as well as the linker histone H1 (Fig. 5.1). The information content of these post-translational modifications is summarized as the "histone code": post-translational modifications of histones either directly affect chromatin density and accessibility or serve as binding sites for effector proteins, such chromatin modifiers (Chap. 6) or chromatin remodeling complexes (Chap. 7), respectively, and ultimately modulate the initiation and elongation of transcription. This implies that histone modifications are able to store information and determine gene expression. The integration of genome-wide histone modification maps together with patterns of chromatin accessibility, transcription factor binding as well as RNA expression from multiple tissues identified novel relationships between histone modifications and related chromatin structures (Sect. 5.2). Moreover, multiple histone modifications act in a combined way to generate a very specific chromatin structure that determines a specific expression level for each class of genes (Sect. 5.3).

With the 15 chemical modifications mentioned above, which can occur at more than 130 sites on five canonical histones (Fig. 5.1) and some 30 histone variants (Sect. 3.3), the theoretical number of possible combinations is sheer unlimited. However, for some common histone marks, correlations between their presence and the activity of different genomic elements are well established, such as H3K9me3 and H3K27me3 at inactive or poised promoters, H3K27ac and H3K4me3 on active enhancers and promoters as well as H3K36me3 in transcribed gene bodies, respectively (Sect. 5.2).

Lysine is the most frequently modified amino acid, since it can accommodate a number of different modifications, such as several types of acylations and methylation, and reactions with ubiquitin and ubiquitin-like modifiers. These modifications occur in a mutually exclusive manner, so that specific lysine residues, such as H3K27, serve as hubs for the integration of different signaling pathways. The reversible acetylation of the ε-amino group of lysines was already presented as an example in Sect. 3.2.

The signaling based on histone acetylation involves three classes of proteins. Lysine acetyltransferases (KATs, "writers") add acetyl groups to proteins, KDACs ("erasers") remove acetyl groups from proteins and acetyl-lysine binders ("readers") selectively interact with acetylated proteins. These chromatin modifiers are often called HATs and HDACs, but since their specificity is not restricted to histones (Box 5.1), the more general terms KATs and KDACs are used here. The KAT family (22 members), the KDAC family (18 members) and the bromodomain family (46 members), many of which bind acetyl-lysine, will be discussed in Chap. 6.

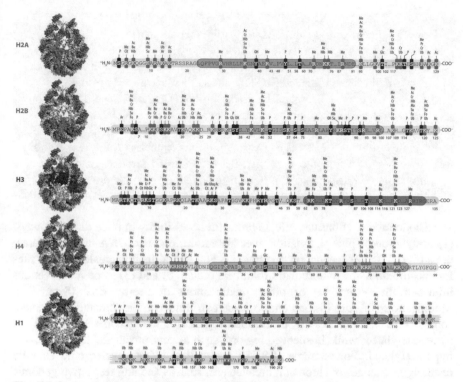

Fig. 5.1 Post-translational modifications of histone proteins. All presently known post-translational modifications of the nucleosome forming histones H2A, H2B, H3 and H4 and the linker histone H1 are indicated. Amino acids that can be modified (K = lysine, R = arginine; S = serine; T = threonine; Y = tyrosine; E = glutamate) are highlighted; most of them can carry different modifications, but not in parallel. Me – methylation (K, R); Ac – acetylation (K, S, T); Pr – propionylation (K); Bu – butyrylation (K); Cr – crotonylation (K); Hib – 2-hydroxyisobutyrylation (K); Ma – malonylation (K); Su – succinylation (K); Fo – formylation (K); Ub – ubiquitination (K); Cit – citrullination (R); Ph – phosphorylation (S, T, Y, H); OH – hydroxylation (Y); Glc – glycation (S, T); Ar – ADP-ribosylation (K, E). For a structural representation of the added group see Fig. 5.4

Box 5.1 Non-histone Protein Modifications

Acetylation of non-histone proteins was described first for the cytoskeletal protein tubulin, the tumor suppressor p53 and the HIV transcriptional regulator Tat. Recent proteomic studies indicated that the majority of acetylation events occur on non-nuclear proteins, such as those found in mitochondria. Acetylation frequently occurs in protein domains, which are structured by α-helices and β-sheets, and is found predominantly on highly conserved proteins, such as metabolic enzymes, ribosomes and chaperones. This is in contrast to phosphorylation sites, which occur often in rapidly evolving,

(continued)

Box 5.1 Non-histone Protein Modifications (continued)

i.e. non-conserved, unstructured protein regions. Non-histone proteins also get methylated. For example, p53 can get methylated at K382 by the writer enzyme KMT5A (also called SETD8) and is then specifically recognized by the tandem Tudor domains of the reader protein TP53BP1. The methylation mark can be removed by the eraser enzyme KDM1A (also known as LSD1), which then prevents the p53-TP53BP1 interaction (for further examples see Fig. 5.7). The PhosphoSitePlus database (http://www.phos phosite.org/homeAction.action) provides access to all presently known post-translation modification sites in both histone and non-histone proteins.

Histone modifications can affect chromatin accessibility in three different ways: (1) acetylation or phosphorylation alter histone-histone and DNA-histone interactions (Fig. 5.2), (2) acetylation, methylation and ubiquitination create binding sites for specific domains of chromatin modifiers (Chap. 6) and (3) some modifications influence the occurrence of other modifications at nearby sites (Sect. 5.3). Modifications of core histone amino acids can influence the structure of nucleosomes (Fig. 5.2). In heavily transcribed regions of the genome histone H3 is hyper-acetylated, while in silenced regions, such as centromers and telomeres, it is hypo-acetylated. For example, H3K56 acetylation affects nucleosome stability resulting in changes in chromatin architecture, such as keeping respective genomic regions more accessible for the binding of transcription factors and Pol II. However, in general the exact position of an acetyl group within a histone is less critical than the site of methylation.

The lysine side chain cannot only be modified by acetylation (two carbon atoms, C2) but also by several other short-chain acyl groups, such as formylation (C1), propionylation (C3), malonylation (C3), crotonylation (C4), butyrylation (C4), succinylation (C4) and glutarylation (C5) and myristoylation (C14) (Fig. 5.3). Propionyl, butyryl and crotonyl modifications have extended hydrocarbon chains that – in comparison to acetylation – increase the hydrophobic state as well as the space requirement of the lysine residue. In contrast, the malonyl, succinyl and glutaryl modifications are most different from acetylations, since they contain an acidic group that changes the charge at the lysine residue from +1 to −1 (at physiological pH). Finally, the acidic acyl group is most different from the other groups due to its negative charge.

For the addition of all these modifications no specific KAT enzyme has yet been assigned, but it can be assumed that KATs are able to catalyze these reactions. In contrast, several members of the SIRT family (Chap. 6) were shown to remove different types of acyl groups. The functional impact of histone acylations has been described only for a few instances. For example, histones close to promoter and enhancer regions are crotonylated, the level of which increases in maturating spermatogonia, i.e. in a period when in these cells histones are exchanged with protamines (Sect. 8.3).

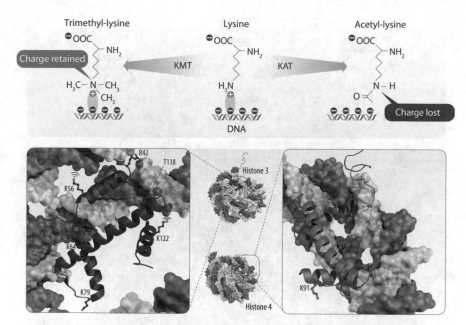

Fig. 5.2 Nucleosome stability through histone modifications. Unmodified lysine residues are positively charged and can form a salt bridge with negatively charged genomic DNA (both at physiological pH). The acetylation of lysines by KATs introduces a bulkier side chain and in parallel removes the positive charge (*top*). This decreases the affinity between DNA and the nucleosome and may destabilize the latter. The methylation of lysines by KMTs does not change the charge but – dependent on the number of added methyl groups – introduces various degrees of bulkiness. Crystal structures of the nucleosome are shown with highlighted key amino acids (*bottom*)

In addition to acylations at lysines further modifications to histones and other proteins are phosphorylations at tyrosines, serines, histidines and threonines, methylations at lysines and arginines, ADP ribosylations at lysines and glutamates, citrullinations of arginines, hydroxylations of tyrosines, glycations of serines and threonines as well as sumoylations and ubiquitinations of lysines (Fig. 5.4).

Methylation is a special type of post-translational modification, since (1) it can occur not only at lysines and arginines, but also at histidines, cysteines, methionines, glutamines, glutamates and asparagines, (2) the methyl group is small and only contributes in a minor way to the steric properties of the amino acid side chains, (3) the methylation of lysines and arginines does not affect the charge of these residues, i.e. also in their methylated form they are positively charged, (4) lysines can be methylated up to three times and arginines up to two times and (5) histone methylations are more stable modifications than phosphorylations or acetylations, i.e. their turnover is lower and they mark more stable epigenetic states.

Examples for the impact of histone methylations with repressive (H3K3me3 and H3K27me3) or activating (H3K4me3) functions have already been discussed in Sects. 1.3, 1.4, 3.4 and 4.2. KMTs transfer one, two or three methyl groups from SAM (Sect. 13.3) to the ε-amino group of lysines, and protein arginine

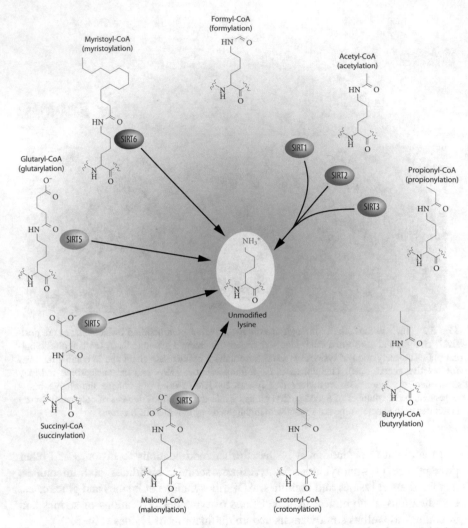

Fig. 5.3 Lysine acylations. The side chain of lysine cannot only be acetylated (*top right*) but also be modified with formyl, propionyl, butyryl, crotonyl, malonyl, succinyl, glutaryl and myristoyl groups, by the respective SIRTs. The various acyl groups differ in the length of their carbon chain, structure and charge states. The members of the SIRT family differ in their specificity for the indicated deacylation reactions

methyltransferases (PRMTs) also use SAM as a methyl donor to add either one or two methyl groups to the guanidinyl group of arginines (Fig. 5.5). Thus, KMTs and PRMTs function as "writers" and change the surface structure of the respective nucleosomes. This allows their recognition by domains of specific methyl-binding proteins ("readers").

In total, the human genome encodes for some 50 KMTs and 30 KDMs that use both histone and non-histone proteins as substrates (Chap. 6). Moreover, the

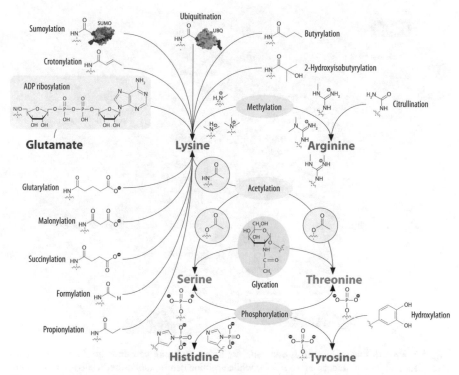

Fig. 5.4 Post-translational modifications of amino acids. The different types of possible post-translational modifications of the amino acids glutamate, lysine, arginine, serine, threonine, histidine and tyrosine are indicated. These modifications do not only occur in histone proteins (Fig. 5.1) but also in many other non-histone proteins (Box 5.1)

human genome encodes nine different PRMTs, but no specific arginine demethylase has yet been identified. The best-known sites of histone methylation are the lysines H3K4, H3K9, H3K27, H3K36, H3K79 and H4K20, while arginine methylation occurs at sites H3R2, H3R8, H3R17, H3R26 and H4R3 (Fig. 5.1). The methylation turnover rates at different lysine residues depend on the function of the respective chromatin mark, but in general they are lower than that of most other post-translational modifications. Thus, methylation is a more specific and long-lasting histone modification than acetylation.

5.2   Genome-Wide Interpretation of Histone Modification Patterns

Genome-wide maps of histone modifications, as obtained by ChIP-seq for different histone markers (Sect. 2.1), were first available from individual research teams and are now part of the huge amount of data provided by ENCODE, Roadmap Epigenomics and IHEC (Sect. 2.2). The aim of these studies was to confirm and

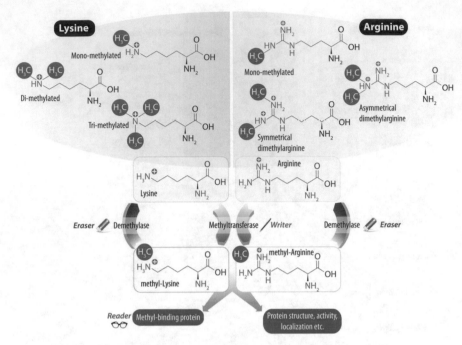

Fig. 5.5 Protein methylation at lysines and arginines. A lysine can undergo mono-, di- or tri-methylation at its ε-amine group (*top left*), while arginine can be mono-methylated or asymmetrically respectively symmetrically di-methylated on its guanidinyl group (*top right*). Methyltransferases ("writers") add the methyl groups, while demethylases ("erasers") remove them (*bottom*). The resulting different protein structures are specifically recognized by respective binding proteins ("readers")

to challenge predictions of the histone code hypothesis (Sect. 5.1) on a genome-wide basis. For example, also on the genome scale – and not only on the single gene level – H3K4me3 and histone acetylation correlate positively with transcription levels. In general, all genome-wide data confirmed that different chromatin regions, such as enhancers, promoters, transcribed genes and heterochromatin, have distinct histone-modification patterns (Table 5.1), which are:

1. Acetylation and deacetylation of histone tails represent major regulatory mechanisms during gene activation and repression (Fig. 1.8). Actively transcribed regions of the genome tend to be hyper-acetylated, whereas inactive regions are hypo-acetylated. The overall degree of acetylation rather than any specific residue is critical.
2. In contrast to acetylation, there is a clear functional distinction between histone methylation marks, both concerning the exact histone residues as well as their degree of modification, such as mono-, di- or tri-methylation. For example, H3K9me3 and H4K20me3 are enriched near boundaries of large heterochromatic domains, while H3K9me1 and H4K20me1 are found primarily in active genes (Figs. 1.3 and 3.6).

Table 5.1 Most conserved histone marks. Non-exclusive list of the relation of post-translational histone marks and their functional impact as confirmed by genome-wide analysis

Histone	Epi-mark	Most conserved co-marks	Status	Genomic region	Biological inference
H3	K4me1	K27me3, K4me2	Poised	Enhancer	
	K4me2	K4me3	Active	Regulatory regions	
	K4me3	K4me2	Active or poised	Promoter	Poised enhancers regulate at many genes as bivalent promoters do
	K9me1		Active	Gene body and enhancer	
	K9me3		Repressive	Heterochromatin at promoters and enhancers	Repetitive sequence (SINEs, LINEs, LTRs)
	K27ac	K4me1/2/3	Active	Enhancer (K4me1/me2)or promoter (K4me2/me3)	H3K27ac marks enhancers and promoters
	K27me1		Active	Enhancer	
	K27me3	K4me1/2/3	Repressive	Polycomb repressed region, promoters	X chromosome inactivation center
	K36me1		Active	Enhancer	
	K36me3	K27ac, K4me1	Active	Gene body	Not correlated with H3K27me3, may be a neglected mark of active enhancers
H4	K20me1		Active	Gene body	
	K20me3		Active or poised	Regulatory regions	

3. H3K4me3 is detected specifically at active promoters, while H3K27me3 is correlated with gene repression over larger genomic regions (Fig. 1.9). Both modifications are usually located in different chromatin domains, but when they coexist on enhancers and/or promoters the respective genomic regions are termed "bivalent." In contrast, latent enhancers are initially not labeled by H3K4me1 or H3K27ac but acquire these active marks and transcription factor binding upon stimulation of cellular signaling pathways.
4. The 5' ends of genes are marked not only by H3K4me2 and H3K4me3, but also by H3 and H4 tail hyper-acetylation (H3K4ac, H3K9ac, H3K14ac, H3K18ac, H3K27ac, H4K5ac, H4K8ac and H4K12ac), H3.3 and H2A.Z variants (Sect. 3.3) and H3K56 acetylation (Fig. 5.2).

5. H3K36me3 levels correlate with levels of gene transcription, since KMTs deposit this mark when interacting with elongating Pol II, i.e. expressed exons have a strong enrichment for this histone mark (Fig. 1.3). Moreover, gene bodies are also marked by H3K79me2/3 and H4K20me1.

6. Histone modification profiles allow the identification of distal enhancer regions, as they show relative H3K4me1 enrichment and H3K4me3 depletion (Fig. 1.3). Interestingly, chromatin patterns at enhancer regions seem to be far more variable, as they show enrichment not only for H3K27ac, but also for H2BK5me1, H3K4me2, H3K9me1, H3K27me1 and H3K36me1, suggesting the redundancy of these histone marks.

7. During the S phase of the cell cycle different areas of the human genome are replicated at different times. Early replicating genes are marked by H3K4me1, 2 & 3, H3K36me3 and H4K20me1 as well as H3K9ac and H3K27ac. In contrast, late replicating genes often correlate with the marks H3K9me2 and H3K9me3. Moreover, boundaries between replicating zones show a pattern of histone signature modifications, such as H3K4me1, 2 & 3, H3K27ac and H3K36me3. This suggests that histone modifications serve as boundary elements, comparable to insulators (see Sect. 4.4) that block spreading of late-replicating heterochromatin.

The non-exclusive list of already understood histone markers (Table 5.1) indicates that genome-wide epigenomic profiling reveals reproducible patterns. These allow predictions on (1) genomic elements that are functional in any given cell type, (2) active genes and (3) silenced genes. In this context, reference epigenomes, as provided by Roadmap Epigenomics and IHEC (Sect. 2.2), are very useful, for example, for distinguishing healthy from diseased cell types.

Genome-wide analysis confirmed that histone marks can differentiate (1) ES cells from differentiated cells and (2) pluripotency genes from lineage-specific genes. In ES cells (Fig. 5.6, *top*), the enhancer regions of both pluripotency genes and lineage-commitment genes are enriched with H3K4me1 and the KAT EP300 (Sect. 6.1). The enrichment of H3K27ac marks activates enhancers of pluripotency genes, whereas the lack of H3K27ac and the presence of H3K27me3 keep lineage-specific genes in a poised state. Accordingly, the promoter regions of pluripotency genes and lineage-specific genes are also active and poised, respectively, exclusively the bodies of pluripotency genes are marked by H3K36me3 and are transcribed.

This demonstrates that the enhancers and promoters of poised genes carry both activating and repressing histone marks, i.e. they are examples of bivalent chromatin states, from which they either get fully activated or repressed. Accordingly, after differentiation towards a specific lineage, such as neurons (Fig. 5.6, *bottom*), only lineage-specific genes are marked by H3K27ac at both enhancer and promoter regions as well as by H3K4me1 at their enhancers. Then Pol II pausing is released allowing mRNA transcription. Genes of other lineages lose marks at their enhancers and obtain H3K27me3 marks at their promoter regions, in order to keep them repressed. Moreover, pluripotency genes attain H3K9me3 marks and DNA methylation at their promoter regions, in order to keep them stably silenced.

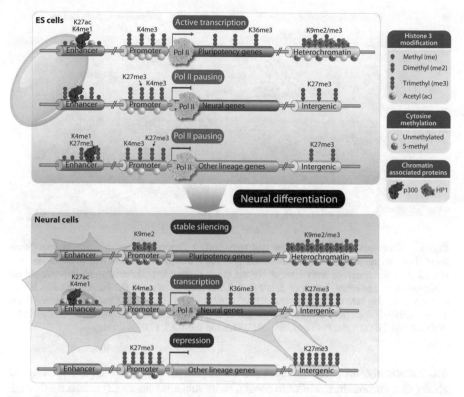

Fig. 5.6 Chromatin states of ES cells in comparison to lineage-specific cells. The chromatin stages at enhancers, promoters, gene bodies and intergenic heterochromatin regions of pluripotent genes, neuronal genes and other lineage genes are compared between ES cells (*top*) and – as an example – neural cells (*bottom*)

During the differentiation process, heterochromatic regions are marked by H3K9me2 and H3K9me3, HP1 binding and DNA methylation are expanded and become more condensed. In repressed genes as well as in intergenic regions H3K27me3 marks also increase.

5.3   Combinatorial Impact of Histone Modifications

The 12 conserved histone marks listed in Table 5.1 already represent the basic "alphabet" of the histone code that is expected to be in future further extended. In addition, the analogy to an alphabet suggests that histone marks may be combined to "words." Only a few histone marks are mutually exclusive. For example, H3R2me2 does not allow H3K4 methylation and phosphorylation of H3S10 prevents H3K9 methylation. Despite these exceptions the remaining histone modifications may be combined to a text rich in information about the local chromatin status.

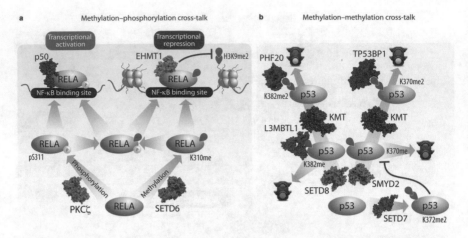

Fig. 5.7 Cross-talk between histones and transcription factors via post-translational modifications. Two examples of the communication between transcription factors and histones are illustrated. The repression or activation of NF-κB target genes via H3K9me2 mark depends on the methylation or phosphorylation of K310 and S311, respectively, of the NF-κB subunit RELA and involves the KMTs SETD6 and EHMT1 (*left*). Similarly, p53-dependent transcriptional activity depends on mono- and di-methylation of K370, K372 and K382 and involves the KMTs SMYD2 and SETD7, the PcG protein L3MBTL1 and the KAT PHF20 (*right*)

Mechanistically, this implies that a pre-existing modification affects (1) directly the ability of a chromatin modifier to recognize its substrate site or (2) indirectly through the recruitment of reader proteins that then recruits the chromatin modifier.

Similarly, histone modifications can be functionally linked with DNA methylation. The example of bimodal methylation (Fig. 4.5) already demonstrated that H3K4me3 marks prevent DNA methylation by disabling the recognition of genomic DNA by DNMT3L. In contrast, H3K9me3 marks on unmethylated CpGs, which are located at TSS regions, recruit DNMT3A and DNMT3B. This results in de novo DNA methylation, i.e. H3K9 methylation is required for DNA methylation.

Furthermore, post-translational modifications of transcription factors interfere with histone marks. Nuclear factor-κB (NF-κB), the key transcription factor in the immune system controlling inflammation via the induction of cytokine gene expression (Chap. 12), provides a key example. When the RELA subunit (also called p65) of NF-κB is methylated at K310 by the KMT SETD6, the KMT EHMT1 is enabled to bind and to set a repressive H3K9me2 mark at the genomic NF-κB binding region (Fig. 5.7, *left*). However, phosphorylation of RELA at S311 by protein kinase Cζ (PKCζ) blocks EHMT1 binding to RELA and relieves the repression of the NF-κB target genes.

The gene of the transcription factor p53, *TP53*, is mutated in more than 50% of all human cancers. In response to various genotoxic stresses, p53 target genes control the processes DNA damage repair, apoptosis and cell cycle arrest (Chap. 10). The activity of the transcription factor p53 is modulated by methylation at two key lysine residues. The mono-methylation of K370 by the KMT SMYD2 represses p53, while di-methylation at the same residue allows binding of TP53BP1 (Box 5.1) and activates p53 target genes (Fig. 5.7, *right*). In addition, demethylation of K372 by the KMT SETD7 inhibits SMYD2-dependent K370 mono-methylation and thus promotes p53-dependent transcriptional activity. Similarly, mono-methylation of K382 enables binding of the repressive PcG protein L3MBTL1 binding, whereas K382 di-methylation allows the interaction with the KAT PHF20 and transcriptional activation.

Key Concepts
- Post-translational modifications of histones are frequent and important epigenetic signals that control many biological processes, such as cellular differentiation in the context of embryogenesis.
- The information content of the post-translational modifications is summarized as the histone code.
- Reversible post-translational modifications, such as phosphorylation, acetylation, and methylation of key amino acid residues are the major mechanisms of communication and information storage of proteins in the control of signaling networks in cells.
- 15 chemical modifications occur at more than 130 sites on five canonical histones and some 30 histone variants.
- Lysine is the most frequently modified amino acid, since it can accommodate a number of different modifications, such as several types of acylations and methylations, and reactions with ubiquitin and ubiquitin-like modifiers.
- The lysine side chain can also be modified by several other short-chain acyl groups, such as formylation (C1), propionylation (C3), malonylation (C3), crotonylation (C4), butyrylation (C4), succinylation (C4), glutarylation (C5) and myristoylation (C14).
- Acetylation and deacetylation of histone tails represent major regulatory mechanisms during gene activation and repression, respectively.
- The signaling based on histone acetylation involves KATs (writers), KDACs (erasers) and acetyl-lysine binders (readers).
- KMTs and PRMTs act as writers and change the surface structure of nucleosomes, thus allowing their recognition by domains of specific methyl-binding proteins (readers).
- The collection of histone modification patterns from various human cell lines and primary tissues, as provided by ENCODE, Roadmap Epigenomics and IHEC, provide a genome-wide basis of the histone code and its impact.
- The marks H3K9me3 and H3K27me3 are found at inactive or poised promoters, H3K27ac and H3K4me3 on active enhancers and promoters, and H3K36me3 in transcribed gene bodies, respectively.
- There is a clear functional distinction between histone methylation marks, both concerning the exact histone residues as well as their degree of modification, such as mono-, di- or tri-methylation.

Additional Reading

Biggar KK, Li SS (2015) Non-histone protein methylation as a regulator of cellular signalling and function. Nat Rev Mol Cell Biol 16:5–17

Carlberg, C., and Molnár, F. (2016). Mechanisms of Gene Regulation. Dordrecht: Springer Textbook. ISBN: 978-94-007-7904-4

Choudhary C, Weinert BT, Nishida Y et al (2014) The growing landscape of lysine acetylation links metabolism and cell signalling. Nat Rev Mol Cell Biol 15:536–550

Sabari BR, Zhang D, Allis CD et al (2017) Metabolic regulation of gene expression through histone acylations. Nat Rev Mol Cell Biol 18:90–101

Tessarz P, Kouzarides T (2014) Histone core modifications regulating nucleosome structure and dynamics. Nat Rev Mol Cell Biol 15:703–708

Chapter 6
Chromatin Modifiers

Abstract The activity of chromatin is modulated by a group of enzymes, which catalyze rather minor changes in histone proteins. The human genome expresses in a tissue-specific fashion hundreds of these chromatin modifiers that recognize ("read"), add ("write") and remove ("erase") post-translational histone markers. Most chromatin modifiers are components of larger protein complexes that use bromodomains, chromodomains, PHD fingers or other domains as specific modules for recognizing chromatin marks. Chromatin acetylation is generally associated with transcriptional activation and controlled by two classes of antagonizing chromatin modifiers, KATs and KDACs. In analogy, respectively for histone methylation there are two classes of enzymes with opposite functions, KMTs and KDMs. Since histone methylation can be a repressive as well as an active marker, the exact position in the histone tail and its degree of methylation (mono-, di- or tri-methylation) is critical. The genome-wide view indicates that chromatin modifiers of antagonizing activity frequently co-localize to common genomic loci and mutually fine-tune each other in the control of active, poised and silent genes. The importance of appropriate maintenance of histone modification patterns during cell cycle progression is emphasized by the fact that the dys-regulation of chromatin modifiers, such as KDMs, can lead to cancer.

In this chapter, we will present the specific domains that mediate "reader" function for chromatin modifiers and other nuclear proteins. Moreover, we will discuss the protein families of KATs and KMTs as well as KDACs and KDMs and their function as "writers" and "erasers." We will learn that antagonizing enzymes, such as KATs and KDACs, are often co-localized at chromatin regions where they fine-tune gene expression. Finally, at the example of KDMs, we will emphasize the impact of chromatin modifiers during the control of cell cycle progression.

Keywords Chromatin modifiers · bromodomain · chromodomain · PHD finger · KATs · KDACs · KMTs · KDMs · writers · erasers · readers · antagonizing activity · genome-wide view · cell cycle control

6.1   Chromatin Readers

Post-translational histone marks represent a kind of chromatin indexing which many chromatin modifiers are able to read via a small set of common recognition domains (Fig. 6.1):

1. Chromatin-organization modifier domains (chromodomains), such as in PcG members, interact with methylated chromatin.
2. The plant homeodomain (PHD) finger was initially identified as a specific reading motif for H3K4me2 and H3K4me3, but tandem PHD fingers also bind acetylated lysines. The domain is contained in some 15 nuclear proteins, such as the KATs CREBBP (KAT3A), EP300 (KAT3B), MYST3 (KAT6A) and MYST4 (KAT6B), the co-repressor TRIM24 and the BRG1-associated factor (BAF) chromatin remodeling complex subunits DPF1, DPF2 and DPF3.
3. Bromodomains, such as those in KATs, recognize acetylated histones and other proteins. Moreover, the combination of a bromodomain with a PHD domain, for example, in the TRIM24 protein, creates a motif on the same nucleosome that specifically recognizes histone H3K23 acetylation in the absence of H3K4 methylation. Thus, proteins with multiple histone binding domains are well suited for the integration of combinatorial messages contained in the histone code.
4. The YEATS (YAF9, ENL, AF9, TAF14 and SAS5) domain recognizes acetylated lysines and has even higher affinity for crotonylated lysines. The domain

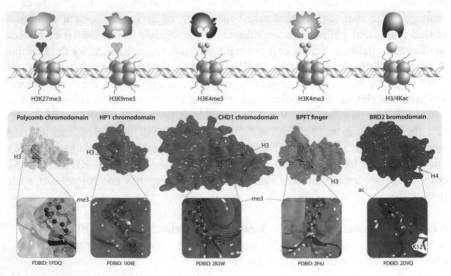

Fig. 6.1 Histone-associated proteins. There are three major types of domains, such as chromodomains, PHD fingers and bromodomains, by which histone modifications are recognized ("read"). While chromodomains and PHD fingers bind to specific histone methylations, bromodomains are rather unspecific and recognize all forms of acetylated histones

is highly conserved and present in proteins that are involved in the regulation of transcription, such as super elongation complex member MLLT3.
5. Tudor domains recognize methylated lysines and arginines.

Lysine acetylation is primarily read by proteins containing a bromodomain. The human proteome encodes 61 bromodomains in 46 different proteins that are found both in the nucleus and in the cytoplasm. These are (1) KATs, such as KAT2A and KAT2B, (2) KAT-associated proteins, such as bromodomain-containing protein (BRD) 9, (3) KMTs, such as KMT2A (MLL1) and KMT2H (ASH1L), (4) helicases, such as members of the SWI/SNF-related matrix-associated actin-dependent regulators of chromatin subfamily A (SMARCA), (5) ATP-dependent chromatin remodeling proteins, such as bromodomain adjacent to zinc-finger domain protein 1B (BAZ1B), (6) transcriptional co-activators, such as TRIMs and TBP-associated factors (TAFs), (7) nuclear scaffolding proteins, such as polybromo 1 (PBRM1), as well as (8) proteins of the bromodomains and extra-terminal domain (BET) family. In contrast, chromodomains are far more specific for a given chromatin modification, i.e. chromodomain-containing nuclear proteins recognize their genomic targets with far higher accuracy than bromodomain proteins.

6.2 Chromatin Writers

Histone acetyltransferases that specifically acetylate lysines are termed KATs, some of which contain bromodomains (Fig. 6.2). They are located in the nucleus and the cytoplasm. Cytoplasmic KATs acetylate histones H3 and H4 post-translationally, which is important for being deposited onto chromatin during DNA replication and repair. KATs use acetyl-CoA as an essential cofactor to donate an acetyl group to the target lysine residue. They can be grouped into three major families: EP300/ CREBBP, GCN5 and MYST. KATs from all three families can also use different acyl-CoAs as substrates for histone lysine acylation. The EP300/CREBBP family members do not only acetylate histones, but also modify basal transcription factors, such as TFIIE, signal-dependent transcription factors, such as p53, and architectural proteins, such as HMGA1. KATs are found at genomic regions that show high levels of histone acetylation, Pol II binding and gene expression. For example, CREBBP (KAT3A) and EP300 (KAT3B) associate both with enhancer and promoter regions, whereas the binding of KAT2B (PCAF), KAT5 (TIP60) and KAT8 (MYST1) is enriched in promoter and transcribed regions of active genes. In this context, EP300 and CREBBP interact with the activation domains of numerous activated transcription factors, such as ligand-activated nuclear receptors or phosphorylated p53.

As already stated before, chromatin acetylation is generally associated with transcriptional activation, whereas the exact residue of the histone tails that is acetylated is not very critical. However, histone methylation, as mediated by KMTs, mainly mediates chromatin repression, but at certain residues, such as H3K4, it results in activation. Therefore, for histone methylation, in contrast to

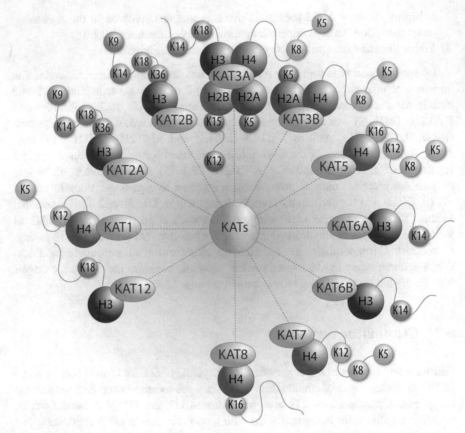

Fig. 6.2 The KAT family. The family of human KATs is a representative example for a family of chromatin modifiers ("writers"). The specific histone substrates of the family members are indicated

acetylation, the exact shape of the residue in the histone tail as well as its degree of methylation (mono-, di- or tri-methylation) is both of critical importance.

KMTs catalyze the transfer of 1, 2 or 3 methyl groups from the donor SAM to the ε-amino group of a lysine residue on a histone, in order to generate mono-, di-, and tri-methylated histones. The human genome encodes 66 KMTs that are subdivided into the following six subfamilies, each of which has a high specificity towards a particular lysine residue:

1. Members of the KMT1 subfamily mediate the mono-, di- and tri-methylation of H3K9. Suppressor of variegation 3–9 homolog (SUV39H) 1 (KMT1A) and SUV39H2 (KMT1B) carry a chromodomain allowing the enzymes not only to write but also to read methylated lysines. Via their MBD domain SETDB1 (KMT1E) and SETDB2 (KMT1F) are able to bind methylated genomic DNA. SETDB1 also has two consecutive Tudor domains and can bind both methy-lated lysines and arginines. Each KMT1 subfamily member has different

affinities to the various methylation degrees of its substrate residue(s). PR/SET domain (PRDM) 2, (KMT8A), PRDM8 (KMT8D), PRDM3 (KMT8E) and PRDM16 (KMT8F) H3K9 are methylases that contain a DNA-binding domain, i.e. they can act both as transcription factor and enzyme. In addition, PRDM3 and PRDM16 act in the cytoplasm as H3K9 mono-methyltransferases and establish there a pool of K9me1-marked H3 histones that in the nucleus, after incorporation to nucleosomes, are converted to H3K9me3.

2. The KMT2 subfamily members, such as Trithorax group (TrxG) family containing MLL proteins, are specialized on H3K4 methylation.
3. KMT3 subfamily proteins restricted on the methylation of H3K36.
4. KMT4 subfamily proteins are specialized on H3K79 methylation.
5. Five KMT5 subfamily members focus on H4K20 methylation: SETD8 (KMT5A), SUV420H1 (KMT5B), SUV420H2 (KMT5C), NSD1 (KMT3B) and ASHL1 (KMT2H).
6. KMT6 subfamily members methylate H3K27. The two main H3K27 methyltransferases are EZH (enhancer of zeste homolog) 1 (KMT6B) and EZH2 (KMT6A), which are catalytic subunits of the PRC2 complex, mediating H3K27 di- and tri-methylation.

KMTs have key roles in regulating cellular processes, such as cell fate determination in response to environmental signals (Sect. 8.4) and use a number of mechanisms in the communication with other chromatin modifiers (Fig. 6.3). In general, H3K9 methylation is associated with transcriptional silencing and heterochromatin formation (Chap. 5). This ensures stable repression at promoters and larger genomic regions. The H3K9-KMTs SUV39H1 and SUV39H2 preferentially act at constitutive heterochromatin. They are gatekeepers of chromatin compaction, since they interact with the linker histone H1, which prevents nucleosome sliding (Sect. 7.1). Moreover, the KMTs recruit chromodomain-containing HP1 proteins, which in turn promote further chromatin condensation.

There are a few mutually reinforcing mechanisms of chromatin silencing:

1. The KMTs SETDB1, EHMT1 and EHMT2 mediate H3K9 methylation within euchromatin and facultative heterochromatin. In this process of dynamic chromatin repression the KMTs often work together with DNMTs, KDACs and H3K4-KDMs, in order to deposit repressing histone marks and remove activating marks at the same time (Fig. 6.3a). This collaboration of KMTs and KDMs is necessary, since the activating mark H3K4me3 prevents H3K9 methylation.
2. Pol II elongates rapidly over constant exons, but at alternatively spliced exons the polymerase needs to be slowed down. At these genomic regions ncRNAs recruit the proteins argonaute 1 (AGO1) and AGO2 and the KMTs SUV39H1, EHMT2 and SETDB1, which creates a repressive chromatin environment through the deposition of H3K9me3 marks and subsequent recruitment of the H3K9me3 reader HP1γ (Fig. 6.3b). This facilitates the recruitment of the spliceosome and thus the modulation of splicing with the result of alternative outcomes.

3. The complex PRC2 interacts with H3K4- and H3K36-KDMs, in order to repress genes through chromatin compaction (Fig. 6.3c). This is another example of coupling the removal of active histone marks with the deposition of repressive marks (here H3K27me3). Although the different grades (mono-, di-, and tri-) of H3K27 methylation are mutually exclusive and are found at different genomic locations, PRC2 controls all H3K27 methylation degrees by initiating their demethylation. H3K27me3 is a docking site for the PRC1 complex via chromobox protein homolog (CBX) proteins, such as CBX7, and results in mono-ubiquitination of H2A (H2AK119ub1) via the RING finger protein BMI1 (BMI1 proto-oncogene, Polycomb ring finger, Sect. 12.1). This promotes chromatin compaction and reduces DNA accessibility for transcription factors and chromatin remodelers (Chap. 7).

4. The PRC2 complex does not only initiate gene repression, but during differentiation it also maintains gene silencing via cooperation with H3K9-KMTs, such as SETDB1, EHMT1 and EHMT2 (Fig. 6.3d).

5. H4K20me3 is enriched at constitutive heterochromatin, but also mediates gene silencing at euchromatic regions (Table 5.1). For example, H4K20me3 deposition by the KMT SMYD5 is associated with silencing of a subset of the target genes of the membrane receptor TLR4 (Toll-like receptor 4) via the

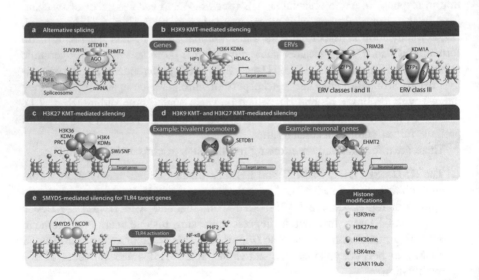

Fig. 6.3 Different mechanisms of KMT-mediated gene silencing. Cooperation of the H3K9-KMTs SETDB1, EHMT1 and EHMT2 with DNMTs, KDACs and H3K4-KDMs in controlling activity of euchromatin and facultative heterochromatin (a). Short-term slowing down of Pol II for the initiation of alternative splicing by the H3K9-KMTs SETDB1, EHMT1 and EHMT2 (b). Interaction of PRC2 with H3K4- and H3K36-KMTs for gene silencing through chromatin compaction (c). Maintaining gene silencing via cooperation of PRC2 with the H3K9-KMTs SETDB1, EHMT1 and EHMT2 in different scenarios of cellular differentiation (d). H4K20me3 deposition by the KMT SMYD5 for silencing TLR4 target genes in co-operation with NCOR1 (e). ERV = endogenous retrovial element

co-repressor protein NCOR (nuclear receptor co-repressor) 1 (Fig. 6.3e). When the bacterial sensing pathway is activated, the transcription factor NF-κB – the mediator of TLR4 signaling – erases H4K20me3 marks from TSS regions via the recruitment of the H4K20-KDM PHF2 (KDM7C).

6.3 Erasers

HDACs (or more specifically KDACs) are highly conserved from yeast to humans. The 18 humans KDACs form four classes: Class I contains HDACs 1, 2, 3 and 8, class II HDACs 4, 5, 6, 7, 9 and 10, class III only HDAC11 and class IV the SIRTs 1-7 (Fig. 6.4). The Zn^{2+}-dependent HDACs 1-11 act predominately in the nucleus and the cytoplasm, while nicotinamide adenine dinucleotide $(NAD)^+$-dependent SIRTs 1-7 are found in addition also in mitochondria. The deacylation

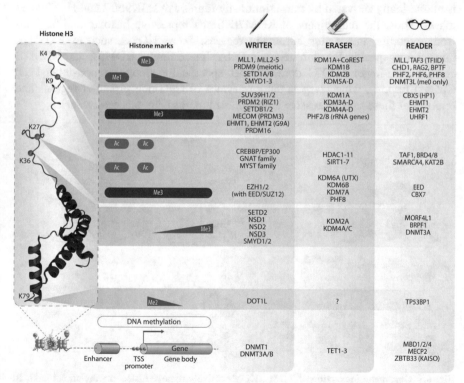

Fig. 6.4 Readers, writers and erasers at histone H3. The interaction with various reader, writer and eraser proteins is indicated through the example of lysines at histone H3. In relation to the elements of a typical gene, i.e. enhancer, promoter and gene body, the distribution of the histone marks is shown as colored bars (*green*: active; *red*: repressive) and wedges indicate their approximate abundance. Histone modification and DNA methylation can cross-talk, as DNMT3A, DNMT3L and UHRF1 contain reader domains for chromatin states

specificities of the SIRT family members were already discussed in Fig. 5.3. KDACs have, like KATs, critical functions in many cellular pathways, and their dys-regulation has been linked to multiple diseases, such as cancer and the metabolic syndrome, and to aging. Accordingly, natural and synthetic compounds that inhibit KDAC activity represent promising drugs (Chaps. 9, 10 and 13).

The human genome encodes 20 KDMs that form two families: (1) the flavin adenine dinucleotide (FAD)-dependent monoamine oxidases LSD1 (KDM1A) and LSD2 (KDM1B) demethylate mono- and di-methylated H3K4 and H3K9 and (2) Fe (II) and α-KG-dependent dioxygenases that contain catalytic Jumonji domains, such as the JARID1 family (KDM5A-D), JHDM1A (KDM2A), JHDM1D (KDM7A) and PHF8 (KDM7B), demethylate mono-, di-, or tri-methylated lysines (Fig. 6.4).

Some KDMs contain reader domains, such as Tudor and PHD domains (Sect. 6.1), that bind to histones and recognize their lysine modifications (Fig. 6.5, *top left*). For example, KDM7 demethylases contain a PHD domain that binds to H3K4me2/3 and recruits these enzymes to genomic regions enriched for this modification. Moreover, KDM7B recognizes H3K4me3 marks of active chromatin and is allosterically activated by transcriptionally repressive H3K9me2 and H4K20me1 marks. Thus, the recruitment of KDM7B limits repressive histone modifications from spreading into active chromatin regions. Some KDMs, such as KDM1A, interact non-specifically with genomic DNA. Interestingly, KDM2A and KDM2B contain zinc-finger CXXC domains that specifically recognize non-methylated CpGs (Chap. 4), which targets these enzymes to CpG-rich promoter regions and removes at these sites repressive H3K36me1/2 marks (Fig. 6.5, *top center*).

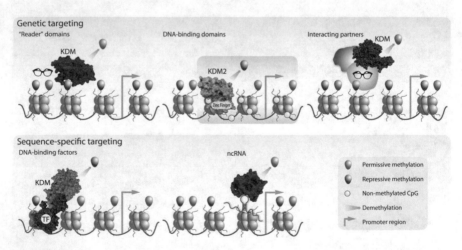

Fig. 6.5 Chromatin interaction of KDMs. KDMs actively remove histone methylation marks, in order to establish at gene regulatory regions new chromatin environments. They recognize their specific genomic target regions by recognizing (1) histone marks via a reader domain (*top left*), (2) unmethylated CpGs via a zinc-finger CXXC domain (*top center*) or (3) histone marks via the reader domain of a partner protein (*top right*). In addition, KDMs can recognize genomic DNA even sequence-specifically via partnering with a transcription factor (*bottom left*) or an ncRNA (*bottom right*)

In addition, in a limited number of cases KDMs can be directed to their chromatin sites by partner proteins that either have a reader domain and recognize chromatin markers (Fig. 6.5, *top right*) or are transcription factors that sequence-specifically bind DNA (Fig. 6.5, *bottom left*). For example, KDM5C interacts with the transcription factor MYC and enriches genome-wide to MYC binding sites. Alternatively, long ncRNAs, such as *HOX* transcript antisense RNA (*HOTAIR*, Sect. 7.4) direct some KDMs, such as KDM1A within the RCOR1 (REST co-repressor) complex, to their chromatin target sites (Fig. 6.5, *bottom right*). RCOR1 is a large protein complex that also contains KDACs and contributes to transcriptional repression. The direction of KDMs via ncRNAs represents a sequence-complementary mechanism for RNA-directed induction of heterochromatin that is also used by the ncRNA *Xist* in the process of X chromosome inactivation (Chap. 4).

6.4 Gene Regulation via Chromatin Modifiers

Cells are constantly exposed to a multitude of signals, such as the extra-cellular matrix, cytokines, peptide hormones and other active compounds, the majority of which derive from receptors at the membrane. This induces signal transduction cascades that often terminate at nuclear proteins, such as transcription factors, chromatin modifying and remodeling proteins, in order to modulate the transcriptome. For example, KDM7C becomes enzymatically active upon phosphorylation by protein kinase A, which allows its interaction with the transcription factor AT-rich interaction domain (ARID) 5B and leads to the recruitment of the KDM to chromatin. Furthermore, phosphorylation of KDM7B by the cyclin E-CDK2 complex regulates gene expression during cell cycle progression (Sect. 6.5).

Most of the signals vary over time and usually have an "on" or "off" character, but the resulting changes in the transcriptome rather have a continuous waveform (Fig. 6.6, *left*). For example, a signal can either directly activate chromatin-modifiers, which then write or erase histone marks, or act indirectly via the activation of nucleosome remodelers that alter the chromatin structure (Chap. 7). In this way, chromatin-associated proteins act as signal converters and integrators. Since the methylation of histones has in general a longer half-life than its acetylation or phosphorylation, the signal can be stored within the epigenomic landscape for shorter and longer time periods, i.e. in particular the histone methylome is suited for a longer-term epigenetic memory. Histone modifications act in combination with DNA methylation and transcription factor activity (Sect. 5.3 and Fig. 6.6, *right*), which increases the diversity of their outputs. Information stored in the epigenome can even be maintained, at least in part, throughout DNA replication (Sect. 9.1).

Genome-wide histone modification maps correspond to different genomic features, such as promoters, enhancers and transcribed genes, or activation states, such as actively transcribed, poised or silenced, and often exist in well-defined combinations (Chap. 5). The mechanistic understanding of the function of antagonizing histone modifiers, such as KATs and KDACs, combined with these

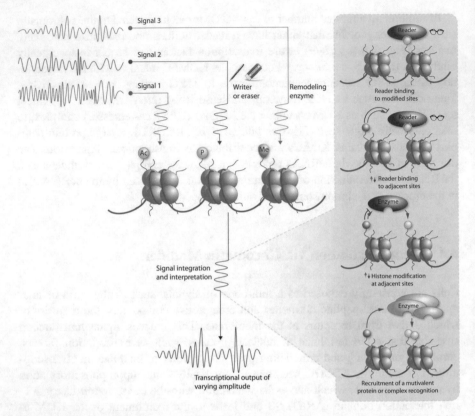

Fig. 6.6 Signal storage and interpretation via chromatin modifiers and readers. Signals deriving from various, mostly membrane-based signal transduction cascades are integrated on chromatin through modifications at histone tails. Multiple inputs occurring over time can be stored. These inputs can affect chromatin directly or are transmitted via chromatin modifiers, such as the writers KATs and KMTs as well as the erasers KDACs and KDMs, and chromatin-remodeling proteins (Chap. 7). There are a number of mechanisms how this dynamic epigenetic landscape is constantly interpreted by reader proteins, such as (1) changing the ability of a reader protein to recognize an adjacent mark, (2) recruiting enzymes that modify additional sites and (3) creating a combinatorial display for recognition in multivalent binding events. The net result of the signal integration can be observed as transcriptional output

genome-wide maps of histone acetylation and methylation results in three main modes for the activity of genes:

1. *Active genes* (Fig. 6.7, *top*): Expressed genes are associated with histone acetylation, H3K4me1, 2 & 3 marks and H2A.Z occurrence in their promoter regions as well as H2BK5me1, H3K9me1, H3K27me1, H3K36me3, H3K79me1, 2 & 3 and H4K20me1 marks in the their transcribed regions. Highest levels of both KATs and KDACs are found associated within these genes and their presence correlates positively with mRNA expression and Pol II levels.
2. *Poised genes* (Fig. 6.7, *center*): These genes are not expressed and do not associate with significant histone acetylation, but they show H3K4 methylation

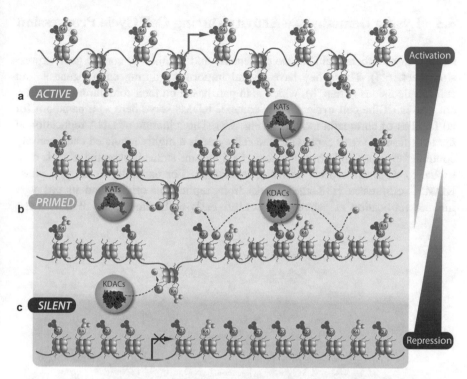

Fig. 6.7 Association of KATs and KDACs with active, poised and silent genes. Both KATs and KDACs are enriched at active genes (*top*). KDACs remove acetyl groups that had been added by KATs after being recruited by elongating Pol II. Lower levels of KATs and KDACs are found at inactive genes being primed by H3K4 methylation (*center*). KDACs prevent Pol II binding and thereby repress transcription via the removal of acetyl groups that had been added by transiently binding KATs. At silent genes that are devoid of H3K4 methylation no KAT or KDAC binding is detectable (*bottom*)

marks and H2A.Z occurrence. There is only low level of KATs or KDACs association with these genes.

3. *Silent genes* (Fig. 6.7, *bottom*): These genes are either associated with H3K27me3 marks together with PcG proteins or are not associated at all with any known chromatin marker. None of these genes are found together with KATs and KDACs.

Taken together, the genome-wide picture suggests that KATs and KDACs work together in modules and associate with the same type of genes. The antagonizing activities of the chromatin modifiers are the basis for precise fine-tuning of gene expression via the homeostasis of active chromatin loci and apply both for promoter and enhancer regions. At primed genes KDACs control the low acetylation level derived from transiently active KATs, in order to prevent Pol II binding. A cycle of transient acetylation and deacetylation keeps these primed genes inactive, but maintains their promoter regions in a poised state waiting for future activation via external signals.

6.5   Lysine Demethylase Activity During Cell Cycle Progression

Chromatin modifiers maintain the epigenome and in this way control gene expression (Sect. 6.4). Thus, they have central importance during embryogenesis and cell fate decisions (Chap. 8), which is in part based on their role during the different phases of the cell cycle. In this context, KDMs serve here as a paradigm for all families of chromatin modifiers (Fig. 6.8). The initiation of DNA replication at the transition of G1 to S phase of the cell cycle is a highly regulated and precisely controlled process, for which the correct epigenome status is essential. KDMs contribute to different aspects of DNA replication. For example, in early S phase KDM5C eliminates H3K4me3 marks from replication origins and in this way initiates replication at actively transcribed early-replicating genes. Furthermore,

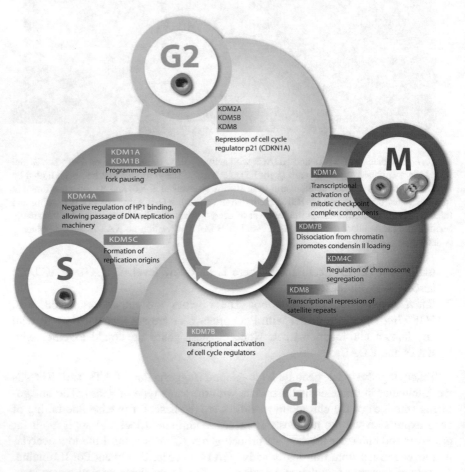

Fig. 6.8 KDM activity during the cell cycle. KDMs play central roles in the indicated processes during the different phases of the cell cycle. Mis-regulation of the activity of these KDMs can cause cell cycle arrest and may lead to genomic instability in cancer

during the S phase KDM4A removes H3K9me3 marks and counteracts HP1γ binding to heterochromatin, in order to allow the passage of the DNA replication machinery. Accordingly, KDM4A protein levels are tightly controlled through the cell cycle. KDM7B removes repressive H3K9me1/2 and H4K20me1 from the promoters of key cell cycle regulators, such as target genes of the transcription factor E2F1, and in this way activates them. This is important for cell cycle progression as well as for G1/S and G2/M transitions. In a similar way, KDM1A activates the expression of components of the mitotic checkpoint complex, which is essential for proper chromosome segregation during mitosis. In parallel, KDM8 stabilizes genomic DNA at repetitive sequences by the elimination of H3K36me2 marks.

The central role of chromatin modifiers in controlling cell cycle progression is also derived from the fact that the mis-regulation of the activity of these proteins can contribute to tumorigenesis (Sect. 10.3). For example, DNA damage response is a process, in which within other proteins also KDM4B and KDM4D are recruited to the sites at which disturbed genomic DNA is sensed. In addition, KDM1A and later also KDM2A are recruited to sites of DNA damage, such as double-strand breaks.

Key Concepts
- Post-translational histone marks represent a kind of chromatin indexing which many chromatin modifiers are able to read via a small set of common recognition domains.
- The human genome expresses, in a tissue-specific fashion, hundreds of these chromatin modifiers that recognize (read), add (write) and remove (erase) post-translational histone markers.
- Most chromatin modifiers are components of larger protein complexes that use bromodomains, chromodomains, PHD fingers or other domains as specific modules for recognizing chromatin marks.
- Chromatin acetylation is generally associated with transcriptional activation and controlled by two classes of antagonizing chromatin modifiers, KATs and KDACs.
- KATs use acetyl-CoA as an essential cofactor to donate an acetyl group to the target lysine residue.
- KATs and KDACs work together in modules and associate with the same type of genes.
- KDACs have, like KATs, critical functions in many cellular pathways, and their dys-regulation has been linked to multiple diseases, such as cancer and the metabolic syndrome, and to aging.
- RCOR1 is a large protein complex that contains KDACs and contributes to transcriptional repression.
- For histone methylation there are two classes of enzymes with opposite functions, KMTs and KDMs.

(continued)

Key Concepts (continued)

- KMTs catalyze the transfer of 1, 2 or 3 methyl groups from the donor SAM to the ε-amino group of a lysine residue on a histone, in order to generate mono-, di-, and tri-methylated histones.
- KMTs have key roles in regulating cellular processes, such as cell fate determination in response to environmental signals, and use a number of mechanisms in the communication with other chromatin modifiers.
- Methylation of histones has in general a longer half-life than their acetylation or phosphorylation, i.e. the histone methylome is suited for a longer-term epigenetic memory.

Additional Reading

Badeaux AI, Shi Y (2013) Emerging roles for chromatin as a signal integration and storage platform. Nat Rev Mol Cell Biol 14:211–224

Carlberg C, Molnár F (2016) Mechanisms of Gene Regulation. Dordrecht: Springer Textbook. ISBN: 978-94-007-7904-4

Dimitrova E, Turberfield AH, Klose RJ (2015) Histone demethylases in chromatin biology and beyond. EMBO Rep 16:1620–1639

Mozzetta C, Boyarchuk E, Pontis J et al (2015) Sound of silence: the properties and functions of repressive Lys methyltransferases. Nat Rev Mol Cell Biol 16:499–513

Chapter 7
Chromatin Remodelers and Organizers

Abstract The accessibility of genomic binding sites for transcription factors and
Pol II at enhancer and promoter regions is essential for efficient gene expression.
Chromatin accessibility is controlled by DNA methylation, histone modifications
and chromatin modifiers, but also chromatin remodeling and 3D chromatin organization are relevant in this process. Chromatin remodelers are multi-protein
complexes that use the energy of ATP hydrolysis, in order to slide, exchange or
remove nucleosomes. Chromatin remodelers make promoter and enhancer regions
either more or less accessible to the transcriptional apparatus, thereby allowing
transcription factors to activate or repress, respectively, the transcription of their
target genes. Thus, nucleosome occupancy and composition are tailored genome-
wide by specialized remodelers. CTCF and cohesin are the main proteins organizing chromatin into architectural loops. Within these TADs, regulatory loops
between enhancer and TSS regions activate their target genes, i.e. the reach of an
enhancer is primarily restricted to the TAD in which it resides. Disease-associated
mutations in chromatin loop anchors and rearrangements of TADs allow a
mechanistic understanding of dys-regulated target genes. Some of the thousands
of long ncRNAs encoded by the human genome are implicated in chromatin regulation and organization. Mechanistic insight into the X chromosome inactivating
ncRNA *Xist* outlines the principles how these molecules regulate gene expression
by coordinating regulatory proteins, localizing to target loci and shaping 3D
chromatin organization.

In this chapter, we will discuss the impact of chromatin remodeling and nucleosome positioning on gene expression. We will present how the 3D structure of
chromatin is organized via CTCF and cohesin, as well as via long ncRNAs and
will learn about their role in disease.

Keywords ATP-dependent remodeling complex · nucleosome dynamics · nuclear
architecture · CTCF · cohesin · TADs · chromatin loops · limb malformations ·
long ncRNAs, *Xist*

7.1   Gene Regulation in the Context of Chromatin Structure

As already discussed in Chaps. 4–6, the access of genomic DNA, in particular at enhancer and promoter regions, is highly restricted by chromatin (summarized in Fig. 7.1). DNA methylation and histone marks store the principal information on DNA accessibility at thousands of local sites within the epigenome of a given cell type. In concert with pioneer transcription factors and other transcriptional co-factors, ATP-dependent chromatin remodelers (Sect. 7.2) open transcription factor binding sites within enhancer regions and release them from steric inhibition by nucleosomes. In addition, DNA looping between enhancer and promoter regions as well as the overall organization of chromatin into functional TAD regions, as mediated primarily by CTCF and cohesin (Sect. 7.3), are essential. Only in this way the activated, enhancer-bound transcription factors get into the proximity of the basal tran-scriptional machinery (containing the Mediator complex, TAFs and Pol II), which are assembled on TSS regions of genes, that should be expressed. The latter is further modulated by ncRNAs, such as eRNAs (Box 2.2) and long ncRNAs (Sect. 7.4). However, all proteins and ncRNAs involved in the control of gene expression work in a mutually interdependent fashion, i.e. there is no fixed sequence how they interact. Taken together, mRNA expression is tightly controlled on multiple levels of chromatin structure.

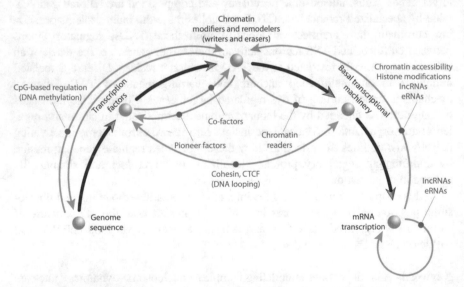

Fig. 7.1 Gene regulation in the context of chromatin structure. The access of transcription fac-tors and Pol II to enhancer and TSS regions is highly restricted by local and global chromatin structures. Methylation of genomic DNA (preferentially at CpGs, Chap. 4) and post-translational modifications of histones (Chaps. 5 and 6) control chromatin accessibility. In addition, chromatin remodelers remove, exchange and evict nucleosomes at chromatin regions that in a given cell type should be recognized by transcription factors and the basal transcriptional machinery con-taining Pol II. This process is supported by pioneer factors, transcriptional co-factors, eRNAs and long ncRNAs. CTCF and cohesin are the main proteins mediating DNA looping, in order to allow contacts between transcription factor-bound enhancers and open TSS regions. At the latter sites gene transcription is initiated

7.2 Chromatin Remodelers

Nucleosomes often block the access of transcription factors to their genomic binding loci, since the packing of genomic DNA around histone octamers hides one side of the DNA. Thus, binding sites that are located close to the center of the 147 bp of genomic DNA, which are covered by a nucleosome, are generally inaccessible to transcription factors. Sites closer to the edge of the nucleosome-covered sequence are a bit better reachable, but only within the linker between nucleosomes genomic DNA is fully accessible. The inaccessibility of a larger part of the genome makes gene expression dependent on chromatin remodeling complexes (Box 7.1) that alter nucleosome position, presence and structure.

Box 7.1 Families of Human Chromatin Remodelers
The names of most remodelers are derived from the nomenclature in yeast, where these complexes were characterized. Due to the high evolutionary conservation of the components of the chromatin remodeling complexes, for human proteins often the name of their yeast homolog is used. The complexes differ in the number of subunits ranging from 4 (ISWI) to 17 (SWI/SNF).

Imitation SWItch (ISWI) family: The ATPase domain of ISWI remodelers contains two RecA-like lobes and a carboxy-terminal HAND-SANT-SLIDE (HSS) domain that binds the unmodified tail of histone H3 and the linker DNA between the nucleosomes. The ISWI complex assembles nucleosomes and spaces them regularly, in order to limit chromatin accessibility and gene expression (Fig. 7.2 *left*). However, nucleosome remodeling factor (NuRF) is a subset of the ISWI complex that promotes transcription.

Chromodomain-helicase-DNA binding (CHD) family: The ATPase domain of CHD remodelers resembles that of ISWI remodelers, but in addition contains at its N-terminus two chromodomains (Sect. 6.1). CHD remodelers conduct assembly (spacing nucleosomes), access (exposing promoters) and editing (incorporating histone H3.3) (Fig. 7.2).

SWItch/Sucrose Non-Fermentable (SWI/SNF) family: The ATPase domain of SWI/SNF remodelers contains two RecA-like lobes and an acting binding N-terminal helicase/SANT-associated (HSA) domain as well as a C-terminal bromodomain (Sect. 6.1). SWI/SNF complexes can use different core paralogs resulting in tissue- and developmental-specific subtypes. The remodelers of this family slide and eject nucleosomes, i.e. they modulate chromatin access (Fig. 7.2, *center*), in order to activate of repress gene expression.

INOsitol requiring (INO) 80 family: These proteins contain a variable, large (> 1,000 amino acids) insertion between the RecA-like lobes. The N-terminus contains an HSA domain that nucleates actin and actin-related proteins (ARPs). Certain INO80 complex subtypes also contain KATs, such as EP400. INO80 remodelers have primarily nucleosome editing functions (Fig. 7.2, *right*), although INO80C also allows chromatin access and nucleosome spacing.

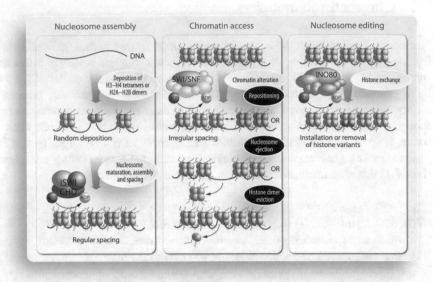

Fig. 7.2 Function of human remodeling enzyme complexes. ISWI and CHD remodelers are involved in the random deposition of histones, the maturation of nucleosomes and their spacing (*left*). SWI/SNF remodelers alter chromatin by repositioning nucleosomes, ejecting octamers or evicting histone dimers (*center*). INO80 remodelers change nucleosome composition by exchanging canonical and variant histones, such as installing H2A.Z variants (*right*). Please note that INO80C, NuRF and certain CHD remodelers have direct transcription activation functions

Since nucleosomes have a rather strong electrostatic attraction for genomic DNA, the catalysis of the sliding, removal or exchange of individual subunits or even the eviction of whole nucleosomes has to dissolve all histone-DNA contacts and requires the investment of energy in form of ATP. The remodeling process involves ATP-dependent DNA translocation, i.e. the dissociation of genomic DNA at the edge of the nucleosome and the formation of a DNA bulge on the histone octamer surface. This DNA loop then moves wave-like over the octamer surface and relocates in this way the DNA without changing the total number of histone-DNA contacts.

In humans, the four families of chromatin remodeling complexes ISWI, CHD, SWI/SNF and INO80 are distinguished based on the differences in their catalytic ATPases and associated subunits. The existence of different complexes suggests that the respective chromatin remodelers have different mechanisms of action. However, all of them contain an ATPase-DNA translocase (Box 7.1). Each remodeler family contains multiple subtypes that provide cell type- or developmentally specific functions. In addition, in humans there exist remodelers that do not belong to one of the four families, such as Cockayne syndrome group B (CSB) and α-thalassemia/mental retardation syndrome X-linked (ATRX). These proteins have major impact on premature aging and mental diseases (Chaps. 9 and 11).

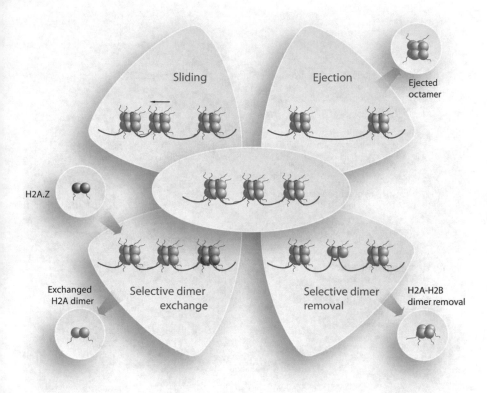

Fig. 7.3 Mobility and stability of nucleosomes. Chromatin remodelers enable access to genomic DNA through sliding, ejection, H2A-H2B dimer removal or selective dimer exchange from nucleosomes. ATP-dependent remodeling complexes as well as thermal motion influence the mobility of nucleosomes. The stability of nucleosomes is affected by its detailed octamer composition and the pattern of histone modifications. For example, the incorporation of histone variants into nucleosomes alters the interactions with histone and non-histone proteins

Chromatin remodelers affect nucleosomes in at least four ways (Fig. 7.3), which are:

1. Movement (sliding) of the histone octamer to a new position within the same chromatin region is the main function of ISWI and CHD remodelers, when it occurs directly after RNA transcription or DNA replication, or by SWI/SNF remodelers, when it provides access to transcription factors as well as to proteins performing DNA repair and recombination.
2. Complete displacement (ejection) of the histone octamer, for example, from promoter regions of heavily expressed "housekeeping" genes, is mainly achieved by SWI/SNF remodelers.
3. Removal of H2A-H2B dimers from the histone octamer by SWI/SNF remodelers leaving only the central H3-H4 tetramer and destabilizing the nucleosome.

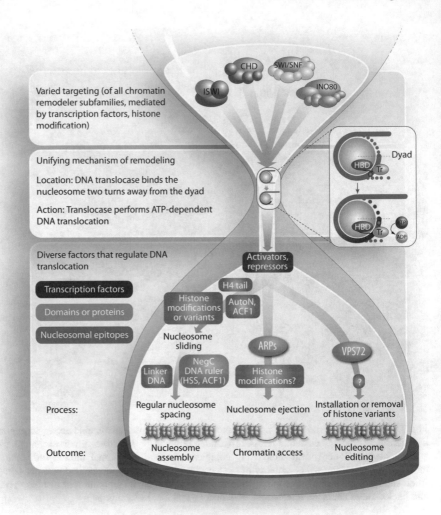

Fig. 7.4 Display of chromatin remodeling. The diversity of the chromatin remodelers and their interaction with transcription factors and histone modifiers (*top*) can explain the variety of outcomes on assembly, ejection and editing of nucleosomes (*bottom*). However, in all cases the core mechanism seems to be the same: the ATPase subunit of each complex is anchored via its histone-binding domain (HBD) two helical turns away from the dyad of the nucleosome core, in order to carry out DNA translocation (*center*). In addition, for each of the different outcomes (*bottom*) particular domains and specific interactions with supporting proteins, such as transcription factors and chromatin marks, are used

4. Replacement of regular histones by their variant forms, such as H2A by H2A.Z (Sect. 3.3). This function is performed primarily by INO80 remodelers and is independent from DNA replication, thus creating a dynamic and transcriptionally permissive state of chromatin.

At steady state, chromatin remodelers ensure dense nucleosome packaging in the vast majority of the genome, while at particular genomic loci they allow the rapid access of transcription factors and other nuclear proteins. The action of SWI/SNF family members is mostly associated with transcriptional activation. Interestingly, the activity of many chromatin remodelers is affected by the presence of histone variants that they themselves introduce into the chromatin, i.e. they control each other's action through the exchange of histones. MacroH2A and H2A.Bbd reduce the efficiency of the SWI/SNF complex, whereas H2A.Z stimulates remodeling by ISWI complexes. The INO80 complex removes H2A.Z from inappropriate locations. In general, H2A.Z resides at open TSS regions and positively regulates gene transcription. The unique amino-terminal tail of this histone variant becomes acetylated when a gene is active.

The ATPases within the different remodeler complexes use a common DNA translocation mechanism (Fig. 7.4), but each complex is formed by a set of different specialized proteins. In this way, the complexes specifically target nucleosomes and differentially detect their epitopes, in order to achieve different outcomes of chromatin remodeling, such as nucleosome assembly, chromatin access and nucleosome editing (Figs. 7.2 and 7.3). For example, chromatin remodelers work together with specific histone chaperones (Box 7.2), in order to ensure spatially and temporally controlled histone deposition and eviction during DNA replication and transcription of genomic DNA.

Box 7.2 Histone Chaperones
There are basically no free histones in a cell, since they are either incorporated into a nucleosome or bound to so-called histone chaperones. The latter are proteins that shield the hydrophobic and charged surfaces of histones, in order to coordinate their controlled transfer onto genomic DNA. During DNA replication, the H3–H4 tetramer is handled by the chaperone proteins ASF1A, ASF1B and CHAF1A, whereas the chaperone NAP1L1 takes care on H2A–H2B dimers. During transcription, the chaperones SUPT6H and SUPT16H remodel nucleosomes before and after passage of Pol II. This means that DNA and RNA polymerases can replicate and transcribe genomic DNA even when it is covered by nucleosomes. During this process the nucleosome does dissociate from the genomic DNA but migrates directly to a more upstream position.

Histone chaperones are involved at the different steps of nucleosome formation, such as (1) binding histones in the cytoplasm directly after their synthesis, in order to regulate their stability, (2) supporting their transport into the nucleus via interaction with the protein importin and (3) affecting the post-translational modification of the histones by facilitating their binding to an appropriate chromatin modifier. For example, histone chaperones deliver H3-H4 tetramers and H2A-H2B dimers to nascent DNA behind the DNA replication machinery, where ISWI and CHD remodelers assemble new nucleosomes. This involves that (1) the initial histone-DNA complexes mature into canonical nucleosomes and (2) the nucleosomes are spaced by relatively fixed distances to each other. This process of assembly and spacing is also used, when during RNA transcription nucleosomes are dynamically ejected.

The tail of histone H4 contains a short basic patch (R17, R18 and R19) that can activate chromatin remodelers, such as ISWI, by orienting the remodeler's ATPase domain to two helical DNA turns away from the dyad of the nucleosome and inducing an allosteric change in their ATP-binding pocket (Fig. 7.4). The N-terminus of the ISWI remodeler contains a mimic of this basic patch that mediates auto-inhibition of the enzyme and is referred to as auto-inhibitory N-terminal (AutoN) domain. The protein ACF1 contributes allosterically to this regulation by hiding away the tail of histone H4. In contrast, CHD remodelers contain two chromodomains in their N-terminus instead of an AutoN domain suggesting a slightly different mechanism. In parallel, the negative regulator of coupling (NegC) domain of ISWI remodelers prevents productive DNA translocation by uncoupling ATP hydrolysis from translocation. Thus, the activity of remodelers can be regulated at the level of their ATPase and their coupling, both of which involve the relief of intrinsic auto-inhibition by epitopes of the nucleosome.

Remodeling complexes contain proteins with bromodomains, chromodomains and PHD domains that can "read" histone marks. For example, ISWI remodelers use a PHD domain for targeting H3K4me3 marked histones, in CHD remodelers chromodomains bind to methylated histones, in SWI/SNF remodelers a bromodomain is used for recognizing acetylated histone H3, while INO80 remodelers employ a bromodomain for binding to acetylated H4. Nucleosome acetylation promotes the recruitment of SWI/SNF remodelers and increases their efficiency. For example, nucleosomes on TSS regions of active genes have a high turnover rate that strongly correlates with H3K56 acetylation. The latter may attract SWI/SNF-family remodelers to eject nucleosomes. This suggests that the activity of both chromatin modifying and remodeling enzymes close to TSS regions increases the turnover of nucleosomes by allowing the inspection of the regulatory genomic region by transcription factors at an increased rate. In contrast, the activities of ISWI and CHD remodelers can be inhibited by histone acetylation. Therefore, ISWI complexes preferentially remodel nucleosomes that lack acetylation, i.e. their activity is focused on transcriptionally inactive regions. In addition, some chromatin modifiers cooperate with remodelers, in particular in the regulation of genes that depend on external signals. For example, CHD complexes contain HDACs 1 and 2, i.e. they also have KDAC activity.

Constitutively active genes, such as housekeeping genes, typically have a nucleosome-depleted region upstream of their TSS, within which key transcription factor binding sites reside. Genome-wide studies indicated that often a 200 bp nucleosome-depleted region upstream from the TSS is flanked on either side by well-positioned nucleosomes. The +1 nucleosome plays a central role in determining the activity of Pol II. At active genes the +1 nucleosome is found approximately 40 bp downstream of the TSS, while at inactive genes the nucleosome is located 10 bp downstream of the TSS. Pol II is frequently stalled at the +1 nucleosome. This stalling is also referred to as "poising" (Sect. 5.2), when transcription is blocked until a signal for activation or release is received, or as "pausing," when Pol II is slowed down immediately downstream of the TSS. Therefore, the +1 nucleosome either physically blocks the progression of Pol II or regulates the presence and/or activity of proteins that support Pol II to overcome the stalling. For example, the +1 nucleosome shows high levels of H3K4me3 marks that are bound by the PHD finger of the TAF3 subunit of the basal transcriptional machinery, i.e. the H3K4me3 marks are also present on promoters with non-elongating Pol II.

7.3   Chromatin Organization in 3D

Genome-wide 3C-based methods, such as 5C, Hi-C and ChIA-PET (Sect. 2.1), mapped the whole human genome for chromatin loops, such as TADs. TADs provide boundaries for self-interacting chromatin and thus organize regulatory landscapes, i.e. they define the genomic regions, in which enhancers can interact with TSS regions of their target gene(s) (Fig. 7.5). In general, the probability that two regions of a chromosome contact each other by chance via DNA looping rapidly decreases with the increase of their linear distance. However, when the contact between the two regions is stabilized, for example, by associated proteins, consecutively architectural loops and regulatory loops are formed. Most architectural

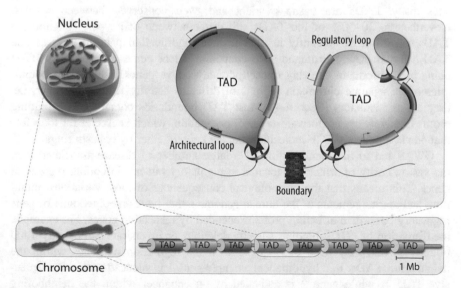

Fig. 7.5 Organization of chromosomes into TADs. The human genome is subdivided into a few thousand TADs defining genomic regions, in which most genes have their specific regulatory elements, such as promoter(s) and enhancer(s). TADs are architectural loops of chromatin that are insulated from each other by anchor regions binding complexes of CTCF and cohesin. Within TADs smaller regulatory loops between enhancers and promoters are formed

loops are identical to TADs (also sometimes referred to as insulated neighbor-hoods), since they are anchored by CTCF-CTCF homodimers in complex with cohesin and carry at least one gene. Lower resolution Hi-C studies initially determined the linear size of TADs in the range of 100 kb to 5 Mb (median 1 Mb, whereas higher resolution studies in human ES cells found 13,000 TADs ranging from 25 to 940 kb (median 190 kb) and containing 1–10 genes (median 3 genes).

Regulatory loops are formed between enhancers and TSS regions that are located within the same TAD, i.e. they are smaller than TADs (Fig. 7.5). In fact, the maximal linear size distance between an enhancer and the TSS of a gene that it is regulating is determined by the size of the TAD within they reside. The for-mation of regulatory loops relies on the binding of transcription factors to the enhancer regions and in a few cases depends also on CTCF and cohesin. The functional result of regulatory loop formation is the stimulation of gene expres-sion. Interestingly, there are also regulatory loops between the beginning and the end of a gene. For example, the promoter regions of rRNA genes form loops with terminator sequences of the same gene. These loops are associated with increased rRNA expression, because they facilitate reloading of RNA polymerase I to the TSS. This promoter-terminator looping applies also for some Pol II genes. Such gene loops represent a kind of transcriptional memory, where a loop formed after an initial round of gene activation speeds up the reactivation of the gene.

TADs subdivide chromosomes into structural domains and also serve as func-tional units (Fig. 7.5). Due to the high evolutionary conservation of CTCF binding sites many TADs are tissue-invariant and even conserved between species. Nevertheless, 30–50% of the TADs still differ between cell types. Neighboring TADs can differ significantly in their histone modification pattern, such as one TAD being in heterochromatic state containing silent genes and the other TAD being in euchromatin carrying transcriptionally active genes. Thus, TAD bound-aries often separate chromatin regions of different activity from each other, i.e. they act as insulators (Sect. 4.4). Most TAD boundaries contain CTCF-binding motifs in a convergent (forward-reverse) orientation, which is crucial for loop for-mation via a pair of CTCF molecules that are kept together by cohesin rings.

GWAS led to the identification of a large number of disease-associated loci, the vast majority of which are located in regulatory but not in coding regions of genes. This means that the physiological consequences of most variations, muta-tions and rearrangements of the human genome rather have an epigenomic or gene regulatory basis than affecting proteins encoded by respective genes. For example, the disruption of CTCF binding sites by deletions or inversions can destroy TAD boundaries and may have consequences on the regulation of genes within the neighboring TADs. In a hypothetical example gene 1 is silenced within the repres-sive TAD 1, while gene 2 is activated by an enhancer within the neighboring TAD 2 (Fig. 7.6a). This wild type gene regulatory scenario can be disturbed by deletion of the boundary region between the two TADs resulting in activation of gene 1 (Fig. 7.6b). Similarly, gene 1 may be activated by inversion (Fig. 7.6c) or duplication (Fig. 7.6d) of the genomic region of the TAD boundary.

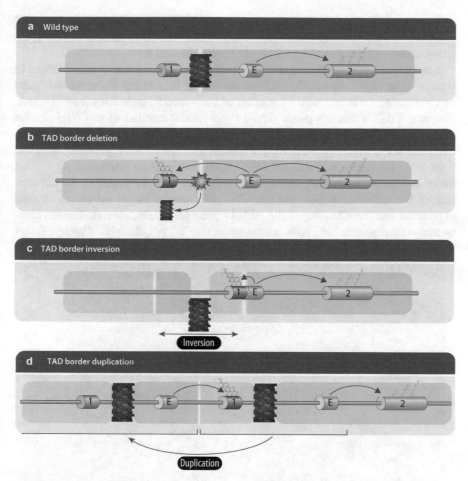

Fig. 7.6 Disturbing TAD structures by genomic mutations or chromosomal rearrangements. In the wild type case TAD 1 (*purple*) is separated by a boundary region ("Brick wall") from TAD 2 (beige) resulting in that gene 1 is insulated from activation by the enhancer (E) in TAD 2 and allows only gene 2 to be active (a). When the TAD boundary is disturbed, for example, by (i) CTCF and/or cohesin depletion, (ii) mutation or (iii) methylation of a CTCF-binding site, TADs 1 and 2 may merge, gene 1 is no longer insulated and can be activated by the enhancer (b). Chromosomal rearrangements, such as inversions (c) or duplications (d) of the boundary region, can merge TADs in a new genomic context resulting in new regulatory scenarios that may result in activation of both genes 1 and 2

Real examples for these scenarios are displayed in Fig. 7.7. In healthy individuals the tumor suppressor gene *GATA2* (GATA binding protein 2) is activated by an enhancer and is located on a different TAD on chromosome 3 than the repressed oncogene *MECOM* (MDS1 and EVI1 complex locus, Fig. 7.7a, *top*). In patients with AML an inversion of this genomic region relocates the enhancer of the *GATA2* gene close to the *MECOM* gene (Fig. 7.7a, *bottom*). The result of this genome rearrangement is the activation of *MECOM* and the silencing of *GATA2* expression. This event is tumor promoting, since MECOM stimulates cell cycle progression, while active GATA2 would have repressed it.

Similar processes are the mechanistic basis of congenital limb malformations in humans (Fig. 7.7b). In healthy individuals the *EPHA4* (EPH Receptor A4) gene on chromosome 2 is activated by a series of limb-specific enhancers located within the same TAD, while the genes *WNT4* (Wnt family member 4) and *IHH* (Indian hedgehog) in the left-flanking TAD and *PAX3* (paired box 3) in the right-flanking TAD are not active (Fig. 7.7b, *top*). A deletion of the *EPHA4* gene including the right TAD boundary activates the *PAX3* gene and leads to brachydactyly ("short finger syndrome"). Similarly, a deletion of the left TAD boundary or the duplication of the enhancer region and the *IHH* gene results in the activation of *IHH* expression and the silencing of the *EPHA4* gene (Fig. 7.7b, *center*). The developmental consequence of both rearrangements is polydactyly ("many finger

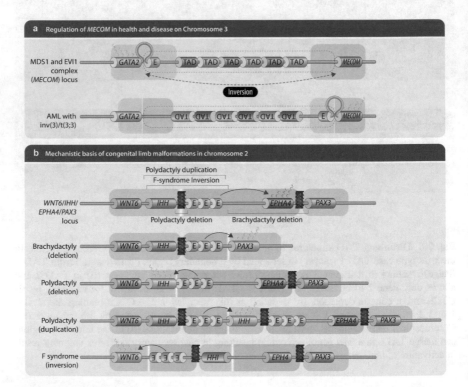

Fig. 7.7 Diseases associated with rearrangements in TADs. In AML, inversions in human chromosome 3 are associated with increased expression of the *MECOM* oncogene (a). This can be explained by a re-direction of an enhancer normally controlling the tumor suppressor gene GATA2 to the *MECOM* gene, with loss and gain of expression of GATA2 and MECOM, respectively. Deletions, inversions and duplications at the locus of the genes *WNT6*, *IHH*, *EPHA4* and *PAX3* on human chromosome 2 in F syndrome disrupt the boundaries of a TAD with limb-specific enhancers ("Stop") (b). Depending on the genes that are under the control of the enhancer this causes the different indicated limb malformations

syndrome"). When the genomic region carrying the *IHH* gene, the left TAD boundary and the enhancers of the *EPHA4* gene are inverted, both genes are silenced, but the *WNT6* gene is activated (Fig. 7.7b, *bottom*). The consequence of this mutation is acropectorovertebral dys-genesis, also referred to as F syndrome, i.e. syndactyly between the first and the second finger and hypodactyly and poly-dactyly of feet. Since the enhancer in the genomic region is limb-specific, all rear-rangements primarily result in malformations of fingers or feet but not of other parts of the body.

7.4   Long ncRNAs as Chromatin Organizers

The long ncRNA (Box 7.3) *Xist* is the key initiator of X chromosome inactivation (XCI) in female cells (Sect. 4.4) and is discussed here as a master example of how ncRNAs contribute to chromatin organization. *Xist* recruits a series of regulatory complexes at different stages of the XCI process and maintains X chromosome-wide transcriptional silencing (Fig. 7.8). In female ES cells both X chromosomes are actively transcribed and carry markers of active chromatin, such as H3K4me1, H3ac and H4ac. However, in early embryonic development, during the blastula stage of approximately 100 cells, XCI is initiated in one of the two X chromosomes by inducing *Xist* expression, which gradually spreads over the whole chromosome. Through the interaction with heterogeneous nuclear ribonucleo-protein U (HNRNPU) *Xist* recruits through its A-repeat region the SMRT/HDAC1-associated repressor protein (SHARP), i.e. SHARP is an RNA-binding protein. Via the co-repressor protein NCOR2 the KDAC HDAC3 is recruited, which leads to demethylation of H3K4 and ejection of Pol II. In addition, *Xist* recruits the complexes PRC1 and PRC2 that deposit H2AK119ub and H3K27me3 marks, respectively. Moreover, the KMT SETDB1 adds repressive H3K9me2 and H3K9me3 marks. In differentiated cells, XCI is maintained by DNA methylation via DNMTs (Chap. 4) and the incorporation of the histone variant macroH2A (Sect. 3.3). In this phase the repressive marks are sufficient for maintaining silen-cing of the X chromosome and *Xist* is dispensable.

Box 7.3 Principles of ncRNA Action

During the past 10 years tens of thousands of RNA transcripts were discov-ered in human tissues and cell types that resemble to mRNAs but do not translated into proteins, i.e. they are ncRNAs. When such an ncRNA is longer than 200 nt, it is called long ncRNA. Due to their broad definition, long ncRNAs are heterogeneous in their biogenesis, abundance and stability, and they differ in the mechanism of action. Some long ncRNA have a clear function, such as in regulation of gene expression, while others, such as eRNAs (Box 2.2), may be primarily side products non-precise of Pol II tran-scription. Many long ncRNAs, such as *Xist*, localize in the nucleus and have

(continued)

Box 7.3 Principles of ncRNA Action (continued)
broad effects on gene expression. They carry out their cellular functions by
interacting with proteins to form macromolecular complexes (Fig. 7.9).
There are specific RNA sequence elements, such as short sequence motifs or
larger secondary or tertiary structures that interact specifically with a large
set of molecular structures in proteins, RNA and DNA. This allows a large
variety of functions, such as the ability to (1) scaffold and recruit multiple
regulatory proteins, (2) localize to specific targets on genomic DNA and (3)
utilize and shape the 3D structure of the nucleus (Fig. 7.9).

During XCI the chromatin of the X chromosome undergoes major structural
changes. Before the expression of *Xist* both X chromosomes are transcriptionally active, not strongly associated with the nuclear lamina and structurally
organized similar to autosomes, i.e. they are subdivided into more than
hundred TADs. However, once *Xist* is expressed by one allele, it spreads
across the X chromosome and interacts with lamin B receptor, which relocates
the chromosome to the nuclear lamina (Fig. 7.10). In this context active
genes are sequestered into the *Xist* compartment and silenced. Moreover, most

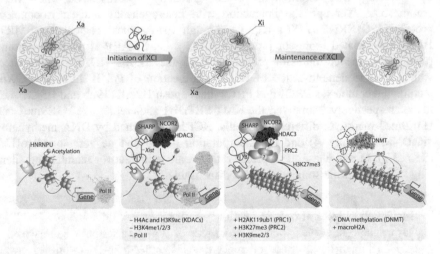

Fig. 7.8 Mechanisms of *Xist*-induced gene silencing. In ES cells both X chromosomes are
actively transcribed (Xa) and are marked by H3K4me1, H3ac and H4ac. XCI starts early in
embryonic development, when *Xist* expression is initiated on one of the two X chromosomes,
and gradually spreads across the whole inactive X chromosome (Xi). *Xist* binds to chromatin
through interactions with HNRNPU and recruits SHARP, in order to promote histone deacetylation via HDAC3, demethylation of H3K4 and ejection of Pol II. Furthermore, *Xist* recruits PRC1
and PCR2 complexes, which deposit H2AK119ub and H3K27me3 marks, respectively.
Moreover, the KMT SETDB1 places repressive H3K9me2 and H3K9me3 marks. XCI is maintained in differentiated cells via DNMT-mediated DNA methylation and the incorporation of the
histone variant macroH2A

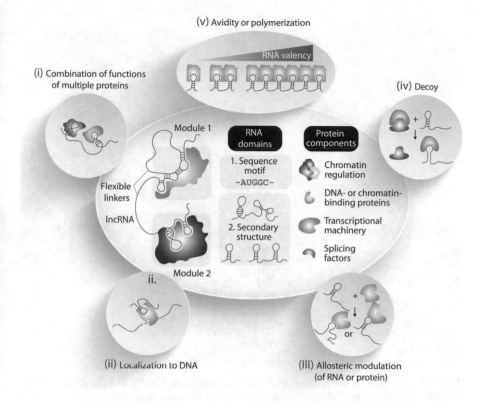

Fig. 7.9 Principles of long ncRNA action. Long ncRNA molecules have various regions for interaction in a modular manner with distinct protein complexes. These interactions have functions, such as (i) combining the functions of multiple proteins, (ii) localizing long ncRNAs to genomic DNA, (iii) modifying the structure of long ncRNAs or proteins, (iv) inhibiting protein function as decoys and (v) providing a multivalent platform, in order to increase the avidity of protein interactions or to promote RNA-protein complex polymerization

of the TADs on the X chromosome are lost and two large mega-domains are formed, which have a boundary at the *DXZ4* locus that associates with the nucleolus.

A number of chromatin modifying and remodeling proteins, such as PRC components, the KMT EHMT2 (KMT1C), the KDM LSD1 (KDM1A), DNMT1 and the SWI/SNF complex, interact with nuclear long ncRNAs. These RNA-protein interactions (1) recruit chromatin regulatory complexes to specific genomic sites, in order to regulate gene expression, (2) competitively or allosterically modulate the function of nuclear proteins and (3) combine and coordinate the functions of independent protein complexes (Fig. 7.9). For example, the long ncRNA *HOTAIR* (Sect. 6.3) associates with PRC2 and the LSD1-RCOR1 complex and erases H3K4me2 marks being associated with gene activation (see also Sect. 11.3).

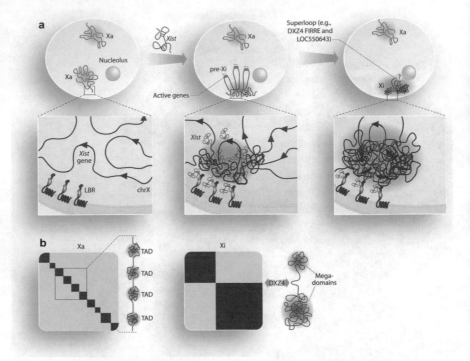

Fig. 7.10 XCI changes the architecture of the X chromosome. During XCI the X chromosome undergoes major structural changes, such as association with nuclear lamina and silencing active genes by sequestering them into the *Xist* compartment (a). The heatmap diagram depicts contact frequency between genomic sites on the X chromosome (b, compare Fig. 3.7)

The abundance of long ncRNAs in cells is related to their function. Low-abundance long ncRNAs, such as *HOTTIP* (HOXA transcript at the distal tip), that in average has less than 1 copy per cell only regulate genes in their close proximity. Moderate expression levels, such as 50–100 copies per cell, enable *Xist* to spread across the entire X chromosome but it does not affect other chromosomes. In contrast, the highly abundant (approximately 3,000 copies per cell) long ncRNA *MALAT1* (metastasis associated lung adenocarcinoma transcript 1) diffuses throughout the whole nucleus and affects many loci. Genome-wide mapping of *MALAT1* binding loci indicates that it associates with all actively transcribed genes in a dynamic and transcription-dependent manner.

Key Concepts
- Chromatin accessibility is controlled by DNA methylation, histone modifications and chromatin modifiers, but also chromatin remodeling and 3D chromatin organization are relevant in this process.

(continued)

Key Concepts (continued)

- The inaccessibility of a larger part of the genome makes gene expression dependent on chromatin remodeling complexes that alter nucleosome position, presence and structure.
- Chromatin remodelers are multi-protein complexes that use the energy of ATP hydrolysis, in order to slide, exchange or remove nucleosomes.
- ISWI and CHD remodelers are involved in the random deposition of histones, the maturation of nucleosomes and their spacing. SWI/SNF remodelers alter chromatin by repositioning nucleosomes, ejecting octamers or evicting histone dimers. INO80 remodelers change nucleosome composition by exchanging canonical and variant histones, such as installing H2A.Z variants.
- ATP-dependent chromatin remodelers, in concert with pioneer transcription factors and other transcriptional co-factors, open transcription factor binding sites within enhancer regions and release them from steric inhibition by nucleosomes.
- Histone chaperones are involved at the different steps of nucleosome formation, such as binding histones in the cytoplasm directly after their synthesis, supporting their transport into the nucleus via interaction with importin, and affecting the post-translational modification of the histones by facilitating their binding to an appropriate chromatin modifier.
- Constitutively active genes, such as housekeeping genes, typically have a nucleosome-depleted region upstream of their TSS, within which key transcription factor binding sites reside.
- The human genome is subdivided into a few thousand TADs defining genomic regions, in which most genes have their specific regulatory elements, such as promoter(s) and enhancer(s).
- Regulatory loops are formed between enhancers and TSS regions that are located within the same TAD, i.e. they are smaller than TADs.
- CTCF and cohesin are the main proteins organizing chromatin into architectural loops, such as TADs and regulatory loops.
- The disruption of CTCF binding sites by deletions or inversions can destroy TAD boundaries and may have consequences on the regulation of genes within the neighboring TADs.
- Disease-associated mutations in chromatin loop anchors and rearrangements of TADs allow a mechanistic understanding of dys-regulated target genes.
- The long ncRNA *Xist* is the key initiator of XCI in female cells and a master example of how ncRNAs contribute to chromatin organization.

Additional Reading

Carlberg C, Molnár F (2016) Mechanisms of Gene Regulation. Dordrecht: Springer Textbook.
    ISBN: 978-94-007-7904-4
Clapier CR, Iwasa J, Cairns BR et al (2017) Mechanisms of action and regulation of ATP-
    dependent chromatin-remodelling complexes. Nat Rev Mol Cell Biol 18:407–422
Engreitz JM, Ollikainen N, Guttman M (2016) Long non-coding RNAs: spatial amplifiers that
    control nuclear structure and gene expression. Nat Rev Mol Cell Biol 17:756–770
Hnisz D, Day DS, Young RA (2016) Insulated neighborhoods: structural and functional units of
    mammalian gene control. Cell 167:1188–1200
Krijger PH, de Laat W (2016) Regulation of disease-associated gene expression in the 3D genome.
    Nat Rev Mol Cell Biol 17:771–782
Narlikar GJ, Sundaramoorthy R, Owen-Hughes T (2013) Mechanisms and functions of ATP-
    dependent chromatin-remodeling enzymes. Cell 154:490–503

Part C
Impact of Epigenomics in Health and Disease

Chapter 8
Embryogenesis and Cellular Differentiation

Abstract Embryonic development is a tightly regulated process that produces from a single zygote the some 400 different tissues and cell types forming the human body. The differentiation program of embryogenesis is under the control of transcription factor networks in the context of significant changes in the epigenetic landscape. A master example of these epigenomic (re)programing processes resulting in changes in DNA methylation and histone modification patterns is the formation and development of PGCs into gametes (oocytes and sperm). The method of cellular reprograming of somatic cells into induced pluripotency by the ectopic expression of the master transcription factors OCT4, SOX2, KLF4 and MYC bears not only the potential to regenerate diseased organs but also provides further insight into the principles of epigenetic control of cellular differentiation. Epigenomic changes and key transcription factors are also involved in the homeostasis of regenerating tissues, such as skin and blood, in which adult stem cells constantly differentiate, in order to restitute lost cells.

In this chapter, we will present the epigenomic principles of (1) early embryonic development, (2) programing of PGCs, (3) induced pluripotency and (4) function of adult stem cells in tissue homeostasis. In this context we will discuss the molecular and cellular similarity between induced pluripotency and tumorigenesis.

Keywords Embryogenesis · ES cells · cell lineage commitment · PGCs · cellular reprograming · induced pluripotency · master transcription factors · gene regulatory networks · adult stem cells · tumorigenesis

8.1 Chromatin Dynamics During Early Human Development

The human body is composed of some 10 trillion cells forming more than 400 different tissues and cell types. Embryogenesis creates this diversity of human cell types based on an identical genome, i.e. cellular diversity is based rather on epigenomics than on genomics. This requires the coordination between (1) an increase in cellular mass and (2) the phenotypic diversification of the expanding cell populations. Thus, embryogenesis is a master example for the critical impact of epigenomics and its

© Springer Nature Singapore Pte Ltd. 2018   123
C. Carlberg, F. Molnár, *Human Epigenomics*, https://doi.org/10.1007/978-981-10-7614-5_8

molecular representative, chromatin, in both human health and disease. In each cell chromatin serves as a specific filter of genomic information and determines which genes are expressed, respectively, which not. Thus, early human development (Box 8.1) is a perfect system for observing the coordination of cell lineage commitment and cell identity specification.

Box 8.1 Early Human Development

In the process of fertilization the two types of haploid gametes, oocyte and sperm, fuse and form the diploid zygote (Fig. 8.1). A series of cleavage divisions of the zygote creates the totipotent 16-cell stage morula. A few days after fertilization cells on the outer part of the morula bind tightly together (a process called compaction) and a cavity forms inside. At the blastocyst stage (50–150 cells) a first differentiation occurs. The outer cells (trophoblasts) are the precursors to extra-embryonic cytotrophoblasts that form chorionic villi and syncytiotrophoblasts, which ingress into the uterus, i.e. these cells form the placenta and extra-embryonic tissues. Even before the blastocyst becomes implanted into the uterine wall, its inner cells, the inner cells mass (ICM), begin to differentiate into two layers, the epiblast and the hypoblast (also known as the primitive endoderm). The epiblast gives rise to some extra-embryonic tissues as well as all to the cells of the later stage embryo and fetus, but the hypoblast is exclusively devoted to making extra-embryonic tissues including the placenta and the yolk sac (Fig. 8.1). Some of the embryonic epiblast cells form PGCs, i.e. the precursors of germ cells, that will later migrate to the gonads (Sect. 8.2). During the gastrulation phase the other cells of the embryonic epiblast turn into the three germ layers ectoderm, mesoderm and endoderm that are the precursors of somatic tissues. The cells of these germ layers are only multipotent, i.e. they cannot differentiate to every other tissue. For example, in a series of sequential differentiation steps ectoderm cells can form epidermis, neural tissue and neural crest, but not kidney (mesoderm-derived) or liver cells (endoderm-derived). Finally, the unipotent progenitor cells produce terminally differentiated cells with specialized functions.

Before fertilization the CpG methylation level of the haploid genomes of sperm and oocyte is 90 and 40%, respectively (Fig. 8.2). The nucleus of sperm is 10-times more condensed than that of somatic cells and most histones are replaced by protamines. As a consequence, the genome of sperm is transcriptionally silent, while in the oocyte many genes are active. After fertilization, at the stage of the one-cell zygote, the two haploid parental genomes initially remain separate in their rather different state of chromatin organization. Then both genomes get extensively demethylated, but the paternal genome much faster than the maternal genome (Fig. 8.2). In parallel, the protamines within the paternal chromatin get exchanged back to canonical histones. In the following cell divisions preceding the blastocyst stage, there is passive demethylation in both parental genomes until the methylation

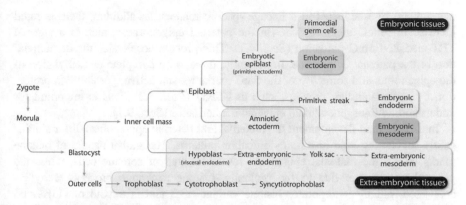

Fig. 8.1 A road map of early human development. Details are provided in the text. The dashed line indicates a possible dual origin of the extra-embryonic mesoderm

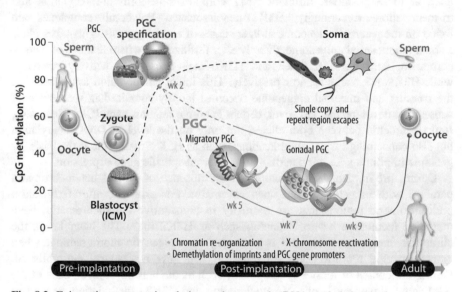

Fig. 8.2 Epigenetic reprograming during embryogenesis. DNA methylation is the most stable epigenetic modification and often mediates permanent gene silencing, during embryogenesis as well as in the adult. Accordingly, there is a hierarchy of events where DNA methylation marks are typically added or removed after changes in histone modifications, i.e. they mostly occur at the end of the differentiation process. Therefore, in this graph the percentage of CpG methylation is indicated as representative marker of epigenome changes. Two waves of global demethylation occur during embryogenesis: the first in the pre-implantation phase (week 1) affecting all cells of the embryo and the other applies only to PGCs during the phase of their specification reaching a minimum of 4.5% CpG methylation from weeks 7 through 9 (Sect. 8.2). Some single copy genomic regions and a number of repeat loci remain methylated and are candidates for transgenerational epigenetic inheritance (Sect. 9.1). Dotted lines indicate the methylation dynamics. Genome-wide de novo DNA methylation takes place after implantation. In PGCs a second wave of de novo DNA methylation happens as early as after week 9 in males, however, as late as after birth in females. Importantly, most CpG promoters are unmethylated at all stages of embryogenesis, i.e. they are not concerned by these waves of methylation and demethylation

patterns are re-established in a lineage-specific manner. In addition, there is rapid increase in 5hmC and 5fC/5caC in the paternal epigenome, which is a sign of TET-mediated 5mC oxidation (Sect. 4.1). This process accelerates the demethylation of the paternal epigenome. Importantly, there is no complete demethylation of the epigenome, and some 5% of the 5mC marks remain active, possibly via protection by methyl-binding proteins, such as ZFP57 (Sect. 4.3). This is important for understanding transgenerational epigenetic inheritance (Sect. 9.1).

In early human development the maternal and paternal epigenomes differ significantly concerning their histone modification patterns. The global pattern of histone marks within the maternal epigenome resembles that of somatic cells, while the paternal epigenome, due to the protamine-histone exchange process, is hyperacetylated, rapidly incorporates histone variant H3.3 and is devoid of H3K9me3 and H3K27me3 markers of constitutive heterochromatin. On the paternal epigenome the first mono-methylation occurs at H3K4, H3K9, H3K27 and H4K20 and later, at these positions, different KMT complexes perform di-methylation and tri-methylation. Accordingly, H3K27me3-associated Polycomb complexes are found on the paternal epigenome at later stages of embryogenesis. This takes place also on the maternal epigenome, directly after fertilization, where features of heterochromatin, such as H4K20me3 and H3K64me3 marks, are actively removed, while H3K9me3 marks are lost passively. This initial "methylation asymmetry" in the paternal and maternal epigenome becomes largely equalized in the course of subsequent development. However, certain genomic regions, such ICRs (Sect. 4.4), stay asymmetric between both alleles, not only on the level of DNA methylation but also concerning histone modifications, such as H3K27me3. This is the basis for genomic imprinting, i.e. for paternal- and maternal-specific gene expression.

During pre-implantation development the absence of typical heterochromatin parallels with in general more open chromatin. For example, after fertilization pericentromeric regions that are normally in constitutive heterochromatin, have marks of facultative heterochromatin, such as H3K27me3. The latter keeps the chromatin more accessible and is necessary for epigenetic reprograming, when gamete-specific modifications are removed and new marks are re-established. During the course of further development, such as the blastocyst level, cells of the ICM show higher levels of DNA methylation and H3K27 methylation, as well as lower levels of histone H2A and/or H4 phosphorylation than cells of the trophectoderm. This epigenomic asymmetry is a sign of differentiation of the respective cell types and regulates lineage allocation in the early embryo. Accordingly, precise and robust gene regulation via the control of access and activity of promoter and enhancer regions is essential (Sect. 1.4).

The chromatin status of TSS and enhancer regions changes significantly on the way from oocytes, the zygote stage, during early embryonic development towards the phylotypic stage, i.e. in the stage when embryos of species display maximal similarity (Fig. 8.3). In oocytes and in the zygotic genome activation (ZGA) stage TSS regions of maternal-zygotic genes are accessible and the respective genes are

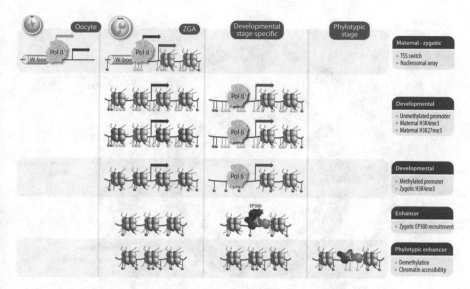

Fig. 8.3 Chromatin dynamics during early embryonic development. The chromatin status of TSS regions (*top and center*) and enhancers (*bottom*) is compared for genes in oocytes, the ZGA stage, early embryogenesis and the phylotypic stage. More details are provided in the text

transcribed. Interestingly, TSS regions that are active in oocytes are AT-rich ("W box," which is similar to a "TATA box"), while those that are active in the zygote are GC-rich. At the zygote stage maternal factors, such as proteins and RNA, via methylation of H3K4 and H3K27 at TSS regions of developmental stage-specific genes, control their expression during early embryogenesis. These genes are often kept silent by H3K27me3 marks as well as the binding of PcG proteins.

During cell lineage commitment, KDMs, such as KDM2A and KDM2B, remove H3K27me3 marks from specific promoter-associated CpGs (Sect. 6.3) in order to make the respective genes transcriptionally permissive. This also includes depletion of nucleosomes from TSS regions (Sect. 7.2). Thus, at different stages of early embryogenesis chromatin dynamics at TSS regions is primarily determined by maternal factors. In contrast, most of developmental stage-specific nuclear proteins, such as the pluripotency transcription factors OCT4 and SOX2 (Sect. 8.3) and the co-activator protein EP300, that are recruited to enhancer regions are newly synthesized and are based on both the maternal and the paternal genome. In addition, enhancers that are specific for the phylotypic stage first need to be demethylated by TET proteins (Sect. 4.1) before transcription factors are able to bind. This suggests that there is a specific role for regulated DNA demethylation during embryonic development. Taken together, chromatin states are changing dynamically during the maternal-zygotic transition and the subsequent embryonic development on the level of histone modification, nucleosome positioning and DNA methylation.

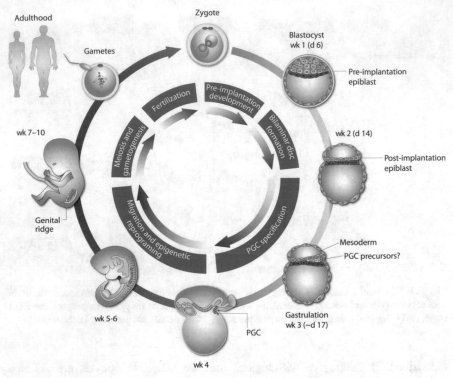

Fig. 8.4 Human germline development. PGCs develop in a cyclic fashion: (1) during fertilization (day zero) a male (sperm) and a female (oocyte) gamete fused to a zygote, (2) during the pre-implantation phase (week 1) the zygote develops into a blastocyst, (3) during the bi-laminar disc formation phase (week 2) pluripotent epiblast cells of the blastocyst give rise to all lineages in the embryo including PGCs, (4) during gastrulation (day 17) the PGC specification phase begins, (5) at week 4 the PGC localize near the yolk sac wall and later migrate to the developing genital ridges, (6) during fetal development and finally adulthood, gonadal germ cells undergo meiosis and gametogenesis, in order to differentiate into sperm and eggs

8.2   Reprograming of the Germ Line

PGCs are the founder cells of the germ line and provide a link between different generations of an individual's family (Chap. 9). The human life starts with fertilization, i.e. the fusion of sperm and oocyte to a zygote. Within the pre-implantation phase (first week) the blastocyst is formed, the cells of which are already differentiated (Fig. 8.4). After implantation into the uterine wall the blastocyst transforms into the bi-laminar embryonic disc, in which ectoderm, mesoderm and endoderm germ layers are formed. During the gastrulation phase (approximately at day 17) precursors of PGCs are specified via gene regulatory networks (Fig. 8.5). First the PGCs are located near the yolk sac wall (week 4) and move then through the hindgut to the developing gonads (week 6). Gonadal PGCs further proliferate until

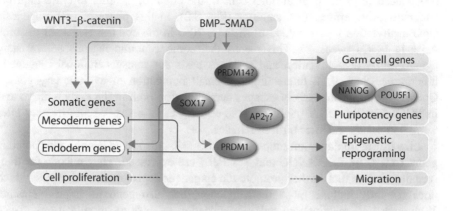

Fig. 8.5 Key transcription factors in PGC specification. The transcription factors SOX17 and PRDM1 form together with TFAP2C and PRDM14 a gene regulatory network that controls PGC specification. SOX17 is upstream of PRDM1, which in turn represses genes that are under the control of SOX17, BMP and WNT signaling. See the text for more details

week 10 and then enter either mitotic quiescence (in male embryos) or the meiotic prophase (in female embryos). In the following fetal development up to adulthood the germ cells undergo meiosis and generate gametes (oocytes or sperm), in the context of which their epigenome again acquires appropriate epigenetic signatures (Fig. 8.2).

PGC specification starts after implantation of the embryo into the uterine wall. Extra-cellular signaling proteins, such as bone morphogenetic proteins (BMPs) and WNTs, which are secreted from extra-embryonic tissues, bind to their respective receptors in the membrane of embryonal cells and induce intra-cellular signal transduction cascades that finally result in the activation of the transcription factors SOX17, PRDM1, PRDM14 and TFAP2C (Fig. 8.5). In this gene regulatory network SOX17 is upstream of PRDM1, which in turn represses SOX17-induced genes in the endoderm as well as BMP- and WNT-induced genes in the mesoderm. Moreover, also PRDM14 and TFAP2C contribute to the germ cell fate. When PGCs migrate to the gonads, the SOX17/PRDM1 gene regulatory network creates a "reset switch" that represses DNA methylation via down-regulation of DNMT3A, DNMT3B and the DNMT1 co-factor UHRF1 (Sect. 4.1), i.e. both DNA methylation maintenance as well as de novo DNA methylation are inhibited in PGCs. In addition, active DNA demethylation by TET1, in particular at ICRs, facilitates the erasure of DNA methylation at parentally imprinted genomic regions. Taken together, both passive and active demethylation mechanisms are used for nearly complete genome-wide erasure of DNA methylation in PGCs.

Genome-wide epigenetic reprograming during pre-implantation development resets the zygotic epigenome for naïve pluripotency. This process is far more pronounced in PGCs than in other cells of the embryo, in order to erase imprints and most other epigenetic memories (Fig. 8.2). In this phase of embryogenesis CpG methylation levels reduce from some 70% in the epiblast (week 2) to some 4.5% in PGCs (week 8). This means that some genomic regions are resistant against DNA demethylation and serve as basis for transgenerational epigenetic inheritance (Sect. 9.1). Interestingly, a small population of evolutionary young, CpG-rich transposons largely escapes demethylation both in the early embryo and in PGCs. Finally, DNA methylation re-establishes in a sex-specific manner after week 9 in males and after birth in females.

Like in pre-implantation embryos and ES cells, also in PGCs the pluripotency genes *OCT4*, *KLF4*, *NANOG* and *TFCP2L1* are expressed. They encode for transcription factors that further promote the epigenetic reprograming of PGCs, which does not only include DNA demethylation, but also genomic imprint erasure, X-chromosome reactivation and reorganization of histone modifications. Since DNA methylation is an important epigenetic silencer that modulates gene expression and maintains genome stability, the transient loss of DNA methylation in PGCs bears the risk causing de-repression of retrotransposons, proliferation defects and even cell death. Therefore, genome-wide reorganization of repressive histone modifications via pluripotency factors safeguards genome integrity in this phase of embryogenesis.

8.3   Cellular Reprograming to Pluripotency

Stem cells have the property of both self-renewal, i.e. the ability to run through numerous cell cycles while staying undifferentiated, and the capacity to differentiate into specialized cell types. Thus, stem cells need to be either totipotent or pluripotent. ES cells are isolated from the ICM of blastocysts, while adult stem cells are found in various tissues (Sect. 8.4). In the embryo, stem cells can differentiate into all types of specialized cells, while in adults they mainly maintain the normal turnover of regenerative organs, such as blood, skin or intestine.

The are different methods to reprogram somatic cells into pluripotency, such as somatic cell nuclear transfer (SCNT) into oocytes, fusion between somatic and pluripotent cells and ectopic expression of defined transcription factors (Box 8.2). These approaches provided important insights into the cellular reprograming process. For example, SCNT showed that epigenetics rather than genetics is the basis for most differentiation processes during normal development. Moreover, cell fusion experiments demonstrated that in cell fusions the pluripotent state is dominant over the somatic state. This led to the discovery that iPS cells can be created by inducing the expression of certain genes, such as those encoding for the transcription factors OCT4, SOX2, KLF4 and MYC, in non-pluripotent cells, like adult somatic cells. Importantly, iPS cells have similar properties as ES cells, including pluripotency and the ability of self-renewal. This means that iPS cells may have the same therapeutic implications and applications as ES cells without the controversial use of embryos in the process of their isolation. However, iPS cells have the potential to be tumorigenic and sometimes show low DNA replication rates and early senescence (Sect. 8.5) (Fig. 8.6).

Box 8.2 Methods for Cellular Reprograming

There are three main methods to transform the state of the nucleus of a non-pluripotent cell into that of a pluripotent early embryonic cell: (1) SCNT, (2) fusion with ES cells and (3) ectopic expression of certain transcription factors (also called direct reprograming). The mechanistic basis of all three approaches is to facilitate the expression of pluripotency-specific genes, such as those encoding for the transcription factors OCT4, SOX2, KLF4 and MYC, which are epigenetically silenced in somatic cells. In SCNT, the genome of somatic nuclei, which were transferred into enucleated oocytes, is remodeled by factors of the ooplasm. Reprogramed SCNT zygotes are totipotent and express OCT4 and other pluripotency factors. In the ES cell fusion method reprograming is initiated in the heterokaryon phase (i.e. both nuclei remain separate) and involves genome-wide epigenetic remodeling including the activation of genomic regions encoding for pluripotency genes. After cell division the somatic and pluripotent nuclei fuse, and additional pluripotency genes are activated. During direct reprograming, OCT4, SOX2, KLF4 and MYC are transfected into somatic cells, which then increase their proliferation rate and have local changes in their epigenome. The transition from a transcription factor-dependent, non-pluripotent state to iPS cells requires multiple cell divisions. However, the iPS cell method is rather robust and the ethically most acceptable way to convert differentiated cells to a pluripotent state.

Direct reprograming of terminally differentiated cells to iPS cells indicates that in principle cell fate is reversible and not necessarily depends on the history of the cell lineage. However, since only 0.1–3% of a cell population get fully reprogramed, i.e. the induction of pluripotency seems to be very inefficient, there seem to be a number of epigenetic barriers that stabilize the identity of somatic cells and prevent their aberrant transdifferentiation.

Many classic pluripotency factors, such as OCT4, SOX2 and NANOG, play a key role in early cell lineage decisions, i.e. they are cell lineage specifiers. Interestingly, in the process of transdifferentiation the expression of a combination of transcription factors in one somatic cell type can directly induce other defined somatic cell fates. For example, the expression of three neuron-specific transcription factors can directly convert fibroblasts into neurons. This implies that theoretically, with the use of appropriate tools, all type of cells are totipotent and may be converted into each other.

In general, reprograming transcription factors have to able to extinct the somatic program and induce a stable pluripotent state, such as of ES cells. Since

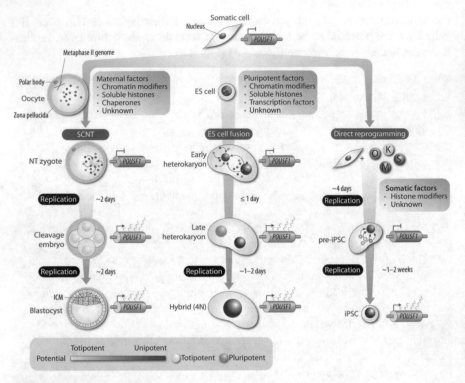

Fig. 8.6 Methods of cellular reprograming. See text for detailed explanations

during normal development from stem cells via progenitor cells to terminally differentiated cells there is a gradual placement of repressive epigenetic marks, such as DNA methylation and histone methylation combined with more restricted accessibility of genomic DNA, reprograming transcription factors need to significantly change the epigenetic stage of somatic cells and to establish that of pluripotent cells. In view of Waddington's model of an epigenetic landscape (Sect. 1.2), in which a "marble" rolling downhill represents development from a stem cell to a terminally differentiated cell (Fig. 1.4), cellular reprograming means to roll back the marble/cell to the top of the hill by epigenomic changes. Moreover, marbles/cells can move only a part of the way up the hill and roll back down passing a different "valley" a discrete number of troughs or even travel from one valley to another without going back uphill (transdifferentiation).

In the presence of serum and the interleukin 6 family cytokine LIF, the most commonly used combination of reprograming transcription factors is OCT4, SOX2, KLF4 and MYC. While OCT4, SOX2 and KLF4 cooperatively suppress lineage-specific genes and activate pluripotency genes, MYC overexpression stimulates cell proliferation, induces a metabolic switch from an oxidative to a glycolytic state and mediates pause release and promoter reloading of Pol II. However, original cell reprograming factors can be replaced either by related transcription

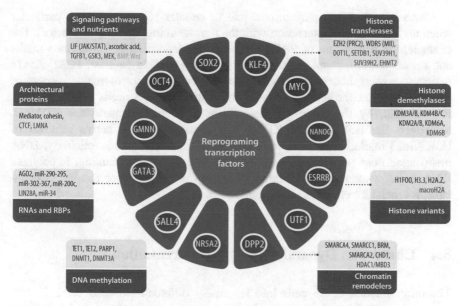

Fig. 8.7 Cell reprograming factors and chromatin state. Different transcription factors that trigger induced pluripotency, such as OCT4, KLF4, SOX2 and MYC, are highlighted. Some molecules have been shown to facilitate (*black*) or inhibit (*red*) cell reprograming. Interestingly, dependent on stage BMPs and WNTs (*blue*) have enhancing or suppressive roles during iPS cell formation

factors, chromatin modifiers, chromatin remodelers, miRNAs or even small compounds (Fig. 8.7). This indicates a remarkable flexibility and redundancy among reprograming factors. For example, OCT4 and SOX2 can be replaced by the pioneer transcription factor GATA3 and the DNA replication inhibitor GMNN.

The obvious first targets of reprograming transcription factors in somatic cells are accessible genomic regions with H3K4me2 and H3K4me3 marks. The following targets will be genomic regions that need to be activated by chromatin remodeling, some of which carry H3K4me1 marks and are in a poised state (Sect. 5.2). The reprograming transcription factors then act as pioneer factors, bind genomic regions that are co-occupied by nucleosomes and recruit other transcription factors and chromatin modifiers (Sect. 4.3). Also bivalent genes with active H3K4me3 marks and repressive H3K27me3 marks belong to this category, but they need to be changed from an active state in somatic cells to a poised state typical for pluripotent cells (Sect. 5.2). During the process of differentiation of ES cells into endoderm, mesoderm and ectoderm lineages most bivalent promoters are converted into monovalent states. Poised enhancers that are marked by H3K4me1 and H3K27me3 are found mostly in ES cells. They often regulate bivalent promoters of genes involved in development. The most difficult targets for OCT4 and similar factors are heterochromatic regions with repressive H3K9me3 marks. These regions need extensive multi-step chromatin remodeling, in order to get transcriptionally activated.

Chromatin modifiers play also a role in cellular reprograming, in particular when they show direct interaction with the reprograming transcription factors. For example, the recruitment of PRC2 results in the deposition of H3K27me3 marks and causes transcriptional repression of somatic genes. In contrast, H3K9-KMTs maintain existing heterochromatin in somatic cells and act as major barriers of reprograming, i.e. the activity of these chromatin modifiers needs to be repressed, for example, by the overexpression of the H3K9me3 demethylase KDM4D. Moreover, the H3K27-KDM KDM6A is required for the active removal of H3K27me3 marks, in order to establish the pluripotent state. Surprisingly, DNA methylation does not play any essential role in cellular reprograming. In contrast, DNA demethylation of pluripotency genes, by either active or passive mechanisms, is crucial for effective reprograming.

8.4   Chromatin Dynamics During Differentiation

The differentiation of ES cells into terminally differentiated cells parallels with progressive transition from open chromatin to a more compact and repressive state. Differentiating cells share accessible chromatin regions with the ES cell they are derived from, but the similarity in the epigenetic landscape decreases when cells mature. After commitment to a specific lineage, the cellular repertoire expands for accessible regulatory regions that contain motifs for transcription factors being specific to that lineage, whereas it clearly decreases for transcription factor binding site of other lineages. Cell type-specific gene activation, via transcription factors binding to enhancers, is a key mechanism for establishing cellular identity. In early embryogenesis, master transcription factors, such as OCT4, SOX2, KLF4 and NANOG, establish and maintain an auto-regulatory transcriptional network, in which they mutually stimulate their expression via different sets of enhancers.

Changes in cell identity are reflected by alterations in the usage of these enhancers. Accordingly, most of the enhancers that are active in early embryogenesis lose their activity during differentiation. This is compensated through the activity of poised enhancers, some of which turn into super-enhancers. These changes in enhancer usage require a chromatin topology that allows a new set of enhancers to interact with their target promoters. In parallel, heterochromatin foci become more condensed and more abundant in differentiated cells than in undifferentiated cells. While in ES cells H3K27me3 marks show only focal distributions, in differentiated cells they largely expanded over silent genes and intergenic regions. This results in (1) silencing of pluripotency genes, (2) activating lineage-specific genes and (3) repressing of lineage-inappropriate genes. Finally, the phenotypes of distinct differentiated cells are gained.

Many adult tissues contain multi-potent stem cells that are able to self-renew and to differentiate into various tissue-specific cell types for lifetime. Thus, adult stem cells are crucial for tissue homeostasis and regeneration. For example, hematopoietic stem cells (HSCs, Sect. 12.1) differentiate into myeloid and lymphoid progenitors that give rise to all cell types of the blood and are responsible for the constant renewal of the tissue. The timing of the expression of master transcription factors plays a key role in these differentiation processes. Moreover, most developmental genes are regulated by multiple enhancers that have both overlapping and distinct spatiotemporal activities. Thus, key lineage genes are associated with dense cluster(s) of highly active enhancers, often referred to as super-enhancers that show stepwise binding of lineage-determining master transcription factors.

Some master transcription factors act as pioneer factors that directly bind nucleosomal DNA to prime enhancers for activation. These pioneers then recruit chromatin remodeling and modifier complexes (Sect. 7.1 and Chap. 5), which in turn facilitate removal and post-translational modification of nucleosomes at these genomic regions. For example, in lymphoid progenitors the sustained expression of the transcription factor CEBPα generates macrophages, whereas under sustained expression of the transcription factor GATA2 mast cells are created (Fig. 8.8). However, when initially CEBPα and afterwards GATA2 are expressed, eosinophils are produced, while the reversed order leads to basophils. The occupancy of these master transcription factors is mostly restricted to genomic regions within accessible chromatin. In HSCs, the marker of active TSS regions, H3K4me3, is more prevalent than differentiated progeny cells, and enhancers of differentiation genes are marked by the mono-methylation of H3K4, H3K9 and H3K27, which are involved in the maintenance of the activation potential that is required for differentiation (Sect. 12.1).

In homeostasis of adult tissues, dividing and differentiating resident stem cells replace damaged or dying cells in the tissue when necessary. Changes in the self-renewal of adult stem cells can either result in premature aging (Sect. 9.3), if it is impaired, or in predisposition to malignant transformation, if it is enhanced (see below). Human epidermis is a master example for stem cell function and differentiation in adult tissues, since it shows an exceptionally high turnover rate. Therefore, the epidermis heavily relies on correct stem cell function, both in homeostasis and after wounding. In the epidermis stem cell pools reside exclusively in the basal layer and proliferation occurs only there (Fig. 8.9). The basal layer continuously replenishes the whole tissue with post-mitotic cells that migrate and differentiate through the different epidermal layers. During this epidermal differentiation process the state of chromatin changes dynamically. Genome-wide levels of H3K27me3 marks decrease during differentiation, in particular at TSS regions of epidermal differentiation genes. This parallels with decreased binding of PRC2 and increased binding of the H3K27me3 demethylase KDM6B. Moreover, in epidermal differentiation the genome-wide histone acetylation level is decreased and terminally differentiated skin cells show higher expression of the KDACs HDAC1 and HDAC2. Finally, also genome-wide DNA methylation levels decrease during epidermal differentiation.

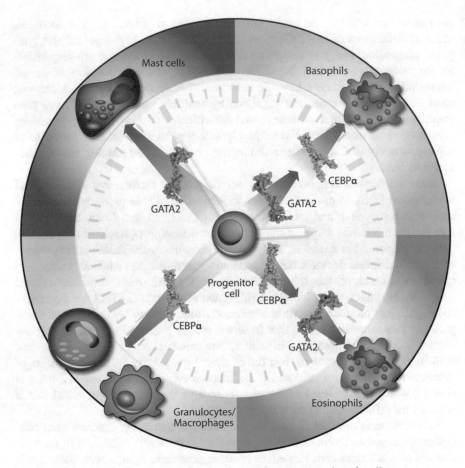

Fig. 8.8 Impact of the timing of transcription factor expression for lineage outcome. Overexpression of CEBPα in common lymphoid progenitors stimulates the formation of macrophages and granulocytes (*bottom left*), whereas high levels of GATA2 induce the formation of mast cells (*top left*). When first CEBPα and then GATA2 are expressed, the cells turn into eosinophils (*bottom right*). Reversing this order of expression leads to basophils (*top right*)

8.5   Development and Disease

Reprograming of a somatic cell to an iPS cell or its transformation to a cancer cell are related events both at the cellular and molecular level. The interpretation of cellular reprograming and transformation as biochemical reactions illustrates that both processes have to overcome a comparable epigenetic barrier that stabilized the starting cells (Fig. 8.10). Both "reactions" are based on multi-step processes that involve proliferation, change the cell identity and finally lead to the formation of immortal cells with tumorigenic potential. Moreover, compared to terminally differentiated cells, adult stem cells or progenitor cells far more likely form either

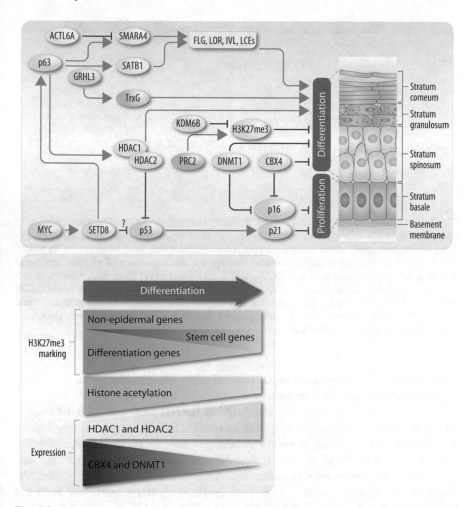

Fig. 8.9 Epigenetics of epidermal stem cells. Proliferation of the epidermis occurs only in the basal layer (*stratum basale, top*). Post-mitotic cells migrate then through the different epidermal layers and differentiate in this way. Finally, they shed from the *stratum corneum*. Epigenetic events that control proliferation as well as lineage commitment and differentiation are shown. During epidermal differentiation the chromatin state is dynamic: H3K27me3 marks are highly locus specific, whereas histone acetylation and DNA methylation generally decrease (*bottom*). Changes in expression levels of key epigenetic regulators are indicated

iPS cells or tumor cells. This suggests that the epigenetic state of stem and progenitor cells is more sensitive to both cellular reprograming and tumorigenesis. In addition, reprograming transcription factors, such as OCT4, SOX2 and KLF4, induce a metabolic switch to create ATP rather from glycolysis than from oxidative phosphorylation, which for cancer cells is known as "Warburg effect." However, iPS cells keep the normal intact diploid genome of the starting cells,

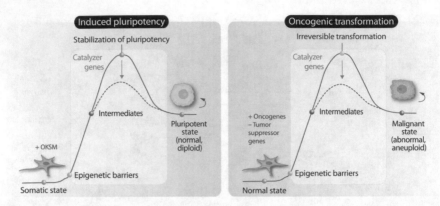

Fig. 8.10 Creation of iPS cells and the risk for malignant transformation. The transformation of somatic cells to iPS cells (cellular reprograming, *left*) or to tumor cells (tumorigenesis, *right*) follows similar principles, where the cells, like in a chemical reaction, have to overcome an activation barrier. Cellular reprograming is started through the overexpression of pluripotency genes encoding for the transcription factors OCT4, SOX2, KLF4 and MYC (Sect. 8.3), whereas in cellular transformation either oncogenes are activated or tumor suppressor genes are repressed. In both cases immortal cells are generated, but iPS cells contain an intact diploid genome, while tumor cells accumulate mutations in their genome that in addition often becomes aneuploid. More details are provided in the text

while cancer cells accumulate mutations and often get aneuploidy, i.e. they have an abnormal number of chromosomes.

All four classic reprograming transcription factors, OCT4, SOX2, KLF4 and MYC, are found to become amplified or mutated in human cancer, i.e. they belong to the some 500 cancer genes known in humans (Sect. 10.3). Moreover, chromatin modifiers, such as KDM2B, that support cellular reprograming (Fig. 8.7) are also associated with tumorigenesis in leukemia and pancreas cancer. In contrast, the expression of histone variants, such as macroH2A, creates a barrier for both iPS cell formation and malignant progression of melanoma cells.

Both cellular reprograming and tumorigenesis show similar genome-wide changes in chromatin structure and DNA methylation. Compared with differentiated cells both cancer and iPS cells have reduced levels of H3K9 methylation and aberrant hyper-methylation or hypo-methylation. For example, reduced methylation levels, such as result of low *DNMT1* expression, can cause T cell lymphomas and promote iPS cell formation (Fig. 8.7). Similarly, *DNMT3A* mutations are found in AML and knockdown of the gene facilitates human iPS cell formation. Thus, cancer cells need to overcome some of the same barriers as iPS cells, in order to alter their cellular states. This implies that cancer cells might even be epigenetically reprogramed into a non-malignant state. This was experimentally proven at the example of melanoma cells, demonstrating that some cancers are not irreversibly locked in a tumorigenic state.

In conclusion, the molecular hallmarks of epigenomics, such as histone modifications and DNA methylation, are important for the identity of normal terminally differentiated cells and can be used for the process of cellular reprograming. Other aspects of epigenetic response are caused, for example, by aging (Sect. 9.3), by metabolic fluctuations of a varying diet (Chap. 13) or by other variations in the environment. Importantly, most of these hallmarks are reversible, such as by the chemical inhibition of chromatin modifiers enzymes, i.e. they may serve as basis for the therapy of cancer, neurological and immunological diseases (Chaps. 10–12).

Key Concepts
- Embryonic development is a tightly regulated process that, originating from a single zygote, produces the some 400 different tissues and cell types forming the human body.
- The differentiation program of embryogenesis is under the control of transcription factor networks, in the context of significant changes in the epigenetic landscape.
- During early embryogenesis there is no complete demethylation of the epigenome and some 5% of the 5mC marks remain active.
- The process of genome-wide epigenetic reprograming during pre-implantation development for resetting the zygotic epigenome for naïve pluripotency is far more pronounced in PGCs than in other embryonic cells, in order to erase imprints and most other epigenetic memories.
- The chromatin status of TSS and enhancer regions changes significantly on the way from oocytes, the zygote stage, during early embryonic development towards the phylotypic stage.
- The method of cellular reprograming of somatic cells into induced pluripotency by the ectopic expression of the master transcription factors OCT4, SOX2, KLF4 and MYC bears not only the potential to regenerate diseased organs, but also provides further insight into the principles of epigenetic control of cellular differentiation.
- The molecular hallmarks of epigenomics, such as histone modifications and DNA methylation, are important for the identity of normal terminally differentiated cells and can be used for the process of cellular reprograming.
- OCT4, SOX2 and KLF4 cooperatively suppress lineage-specific genes and activate pluripotency genes, while MYC overexpression stimulates cell proliferation, induces a metabolic switch from an oxidative to a glycolytic state, and mediates pause release and promoter reloading of Pol II.
- iPS cells may have the same therapeutic implications and applications as ES cells, but have the potential to be tumorigenic, and sometimes show low DNA replication rates and early senescence.
- Epigenomic changes and key transcription factors are also involved in the homeostasis of regenerating tissues, such as skin and blood, in which adult stem cells constantly differentiate, in order to restitute lost cells.
- Key lineage genes are associated with super-enhancers that show step-wise binding of lineage-determining master transcription factors.

(continued)

Key Concepts (continued)
- Some master transcription factors act as pioneer factors that directly bind nucleosomal DNA to prime enhancers for activation.
- Changes in the self-renewal of adult stem cells can either result in premature aging, if it is impaired, or in predisposition to malignant transformation, if it is enhanced.
- The epigenetic state of stem and progenitor cells is more sensitive to both cellular reprograming and tumorigenesis.
- All four classic reprograming transcription factors, OCT4, SOX2, KLF4 and MYC, are found to become amplified or mutated in human cancer, i.e. they belong to the some 500 cancer genes known in humans.
- Both cellular reprograming and tumorigenesis show similar genome-wide changes in chromatin structure and DNA methylation.

Additional Reading

Apostolou E, Hochedlinger K (2013) Chromatin dynamics during cellular reprograming. Nature 502:462–471

Avgustinova A, Benitah SA (2016) Epigenetic control of adult stem cell function. Nat Rev Mol Cell Biol 17:643–658

Burton A, Torres-Padilla ME (2014) Chromatin dynamics in the regulation of cell fate allocation during early embryogenesis. Nat Rev Mol Cell Biol 15:723–734

Li M, Belmonte JC (2017) Ground rules of the pluripotency gene regulatory network. Nat Rev Genet 18:180–191

Perino M, Veenstra GJ (2016) Chromatin control of developmental dynamics and plasticity. Dev Cell 38:610–620

Smith ZD, Sindhu C, Meissner A (2016) Molecular features of cellular reprograming and development. Nat Rev Mol Cell Biol 17:139–154

Srivastava D, DeWitt N (2016) In vivo cellular reprograming: the next generation. Cell 166:1386–1396

Tang WW, Dietmann S, Irie N et al (2015) A unique gene regulatory network resets the human germline epigenome for development. Cell 161:1453–1467

Tang WW, Kobayashi T, Irie N et al (2016) Specification and epigenetic programing of the human germ line. Nat Rev Genet 17:585–600

Chapter 9
Population Epigenomics and Aging

Abstract The epigenome has memory functions, since it is able to preserve the results of cellular perturbations by environmental factors in form of changes in DNA methylation, histone modifications or 3D organization of chromatin. Such epigenetic drifts can be detected in epigenomic patterns, like in DNA methylation maps, that are heritable from parent to daughter cells and may in part even be transferred to the next generation. The concept of transgenerational epigenetic inheritance could explain how lifestyle factors of parents and grandparents, such as daily habits in eating and physical activity, can affect their offspring. Epigenome-wide association studies (EWASs) use genome-wide assays, in order to demonstrate that *cis*- and *trans*-regulatory mechanisms shape patterns of population epigenomic variations in detail and as a whole. Different types of human cohort studies, ideally composed of mono-zygotic twins and best having a longitudinal design, are well suited to identify population epigenomic variations that are associated with human traits, such as disease risk.

Aging may be the most important of these phenotypes, as it is everyone's inevitable result of life. The progressive decline in the function of cells, tissues and organs being associated with aging is affected by both genetic and epigenetic factors, i.e. there are characteristic epigenome-wide changes during aging. Common hallmarks of aging are associated with specific chromatin patterns and chromatin modifiers are able to modulate both life- and healthspan. Thus, epigenomic signatures can serve as biomarkers of aging and may be druggable targets, in order to delay or reverse age-related disease.

In this chapter, we will discuss the molecular basis of epigenetic memory in somatic cells and the potential for transgenerational inheritance via the germ line. We will present different types of human cohort studies that are suited for the investigation of population epigenomics via EWAS and related assays. Finally, we will study the epigenomic basis of the aging process and the use of respective epigenetic signatures as biomarkers and drug targets for improving human healthspan.

Keywords Epigenetic memory · epigenetic drift · transgenerational epigenetic inheritance · EWAS · human cohorts · mono-zygotic twins · aging · epigenetic clock · healthspan

9.1 Epigenetic Memory and Transgenerational Inheritance

Results from the 1,000 Genomes Project (www.internationalgenome.org) indicate that a typical human genome differs in some 4.1–5.0 million sites from the 3.26 Gb of the reference genome, i.e. in not more than 0.15%. Most of these variations are single nucleotide polymorphisms (SNPs), but there are also large structural and copy number variants. The some 100 trillion cells of an individual all carry the same genome with identical SNPs and other variants. With the exception of diseases displaying genome instability, such as cancer, this genome stays stable over the life-span of the person (Fig. 9.1).

In contrast, the some 400 tissues and cell types of an individual significantly differ in their respective epigenomes. In addition, in response to cellular perturbations by diet, stress or other environmental influences these epigenomes vary a lot over time. Although different persons show consistency in the overall epigenome patterns of their tissues, individuals vary far more on the level of their epigenomes than on the level of their genomes (Fig. 9.1). This suggests that phenotypic differences between individuals are rather based on the epigenome than on the genome. For example, a study of the epigenome-wide patterns of five histone markers in 19 individuals indicates that the main inter-individual differences involve chromatin state transitions between active and repressive genomic regions and vice versa. Most variable regions are that carrying bivalent enhancers (H3K4me1 and

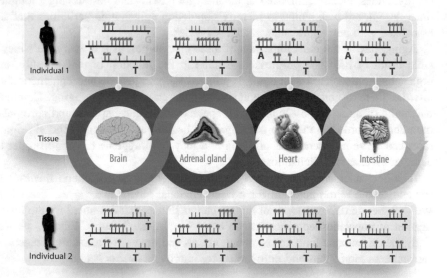

Fig. 9.1 Human individuals show epigenetic heterogeneity. Tissue- and cell type-specific DNA methylation are monitored by clusters of methylated CpGs that vary from organ to organ of the same individual. Filled circles illustrate methylated CpGs and lack of a circle unmethylated CpGs. SNPs are monitored by the corresponding base

H3K27me3 marks), weak enhancers (only H3K4me1 marks) and weak enhancers in transcribed regions (H3K4me1 and H3K36me3 marks).

In principle, any perturbation of cellular homeostasis can result – via epigenetic changes – in long-lasting effects of the phenotype, in particular when the perturbed cells are self-renewing stem/progenitor cells or long-lived, terminally differentiated cells. These perturbations result in the activation of signal transduction pathways that often reach the nucleus. Within the nucleus activated transcription factors communicate with chromatin modifiers and remodelers, in order to create changes in epigenetic signatures, such as histone modifications, DNA methylation and 3D chromatin architecture. Some of these epigenetic changes are very transient (lasting from minutes to hours), while others can stay far longer (days, months or even years).

Both DNA methylation and histone modifications can be transmitted from parent to daughter cells, i.e. epigenetics has a memory function that persists beyond the cell cycle. Accordingly, the epigenome serves as a storage facility of cellular perturbations of the past. This implies that the epigenome is able to memorize lifestyle events in basically every tissue or cell type. Thus, not only neurons store the "memory" of an individual (Chap. 11), but also the immune system memorizes encounters, for example, with microorganisms (Chap. 12), and metabolic organs remember personal habits on diet and physical activity (Chap. 13). In contrast, a disruption of the epigenetic memory can lead to the onset of cancer (Sect. 10.2).

Like in any other biological system with memory, also in epigenetics the state of the system depends on its history. Epigenomic memory provides long-term stability to cell identities via preserved gene expression states and more robust gene regulatory networks. For example, the actions of chromatin modifiers and remodelers within Polycomb and Trithorax complexes (Sect. 6.2) result at hundreds of genomic regions in long-term, mitotically heritable memory of silent and active gene expression states, respectively. Key proteins of these complexes are KMTs and KDMs, such as KMT6A (EZH2), KMT2A (MLL1), KDM5C (JARID1C) and KDM6B (JMJD3), while KATs and KDACs are missing. This re-iterates the principle that short-term "day-to-day" responses of the epigenome are primarily mediated by non-inherited changes in the histone acetylation level (Chap. 5), while long-term decisions, for example, concerning cellular differentiation, are stored in form of histone methylation marks. In addition, DNA methylation is particularly suited for a long-term memory of cells. This means that rather than being a first responder to extra-cellular signals, DNA methylation acts as a consolidator of previously established gene repression in a given cell type (Sect. 4.1).

Interestingly, also long ncRNAs can carry epigenetic memory. A master example is the action of the long ncRNA *Xist* initiating in female cells the silencing of one copy of the X chromosome (Sect. 7.4). Like *Xist*'s association with the X chromosome, the binding of pioneer transcription factors, such as FOXA1, CEBPα or GATA1 (Sect. 3.4), serves as a kind of epigenetic bookmarking and contributes to the epigenetic memory of the respective genomic regions. Similarly, also the binding patterns of key proteins in the 3D chromatin organization, such as CTCF and cohesin (Sect. 3.5), is considered as part of the epigenetic memory.

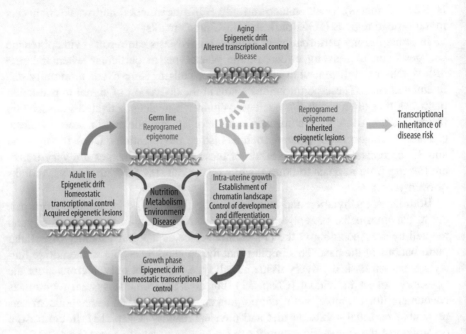

Fig. 9.2 Epigenetic drift and transgenerational inheritance. During embryogenesis epigenetic marks, such as DNA methylation and histone modification, are established, in order to maintain cell lineage commitment (Sect. 8.1). After birth, this chromatin landscape stays dynamic throughout lifespan and responds to nutritional, metabolic, environmental and pathological signals. This epigenetic drift is part of homeostatic adaptations and should keep the individual at good health. However, when an adverse epigenetic drift compromises the capacity of metabolic organs to adequately respond to challenges as provided by nutrition and chronic inflammation, the susceptibility to diseases, such as T2D or cancer, increases. Some of these acquired epigenetic marks can be inherited to subsequent generations when they escape epigenetic reprograming during gametogenesis (Sect. 8.2)

Interestingly, like FOXA1, GATA1 and KTM2A (MLL1), also cohesin remains associated with its genomic binding regions even throughout mitosis.

Epigenetic drifts, such as hyper-methylation of the promoters of tumor promoter genes, contribute to the risk for cancer (Sect. 10.1) and other diseases (Fig. 9.2). In particular the risk for diseases that are caused through exposure with environmental factors, such as microorganisms (chronic inflammation, Sect. 12.2) or food (T2D, Sect. 13.4), showed to have a large epigenetic contribution. Of special interest are diseases that have their onset long time before the phenotype emerges, i.e. where accumulations of epigenetic changes stepwise increase disease susceptibility.

Epigenomic patterns, such as genome-wide DNA methylation, can be stably inherited from mother to daughter cells, i.e. the epigenomic memory is preserved through cell division. Thus, epigenomic information may also be transmitted via the germ line to the next generation (Fig. 9.2). Although some 95% of DNA methylation marks are erased throughout the two rounds of demethylation during PGC generation (Sect. 8.2), some single-copy genomic regions, as well as many retrotransposon loci, such as SINE-VNTR-Alu (SVA) sites (Box 4.2), escape both demethylation waves and remain methylated in gametes. In case the DNA methylation pattern at these escape regions is susceptible to environmental influences, such as dietary molecules (Sect. 13.1), lifestyle experiences of an individual may be transmitted to subsequent generations causing phenotypic consequences. In mice, the transgenerational inheritance of environment-induced metabolic and behavioral traits has already been demonstrated. For example, male mice exposed to in utero undernutrition beget offspring that shows obesity and glucose intolerance. Another master example is that of agouti viable yellow (A^{vy}) mice that carry an IAP retrotransposon close to the gene encoding for the Agouti signaling protein (*Asip*) that mediates inheritable environmental influences on both fur color and BMI.

These rodent models impose the question whether the concept of a metabolic memory is also valid in humans. There are no comparable natural human mutants, and for ethical reasons human embryonal feeding experiments are not possible. However, there are natural experiments, such as followed by the *Dutch Hunger Winter Families Study* (Box 9.1), where individuals were exposed in utero to an extreme undernutrition occurring in the Netherlands during the winter of 1944/45. Accordingly, they had low birth weight but later in life turned overweight and displayed increased incidence of insulin resistance (Sect. 13.4).

In general, the offspring of human mothers being exposed to an adverse environment during embryonic development, such as undernutrition or placental dysfunction leading to impaired blood flow, nutrient transport or hypoxia, during adulthood has an increased risk to develop during adulthood symptoms of the metabolic syndrome, such as obesity, impaired glucose tolerance and finally T2D. These observations were the basis for the "thrifty phenotype" hypothesis (or formulated more general as "developmental origins of health and disease" (DOHaD) hypothesis) suggesting that poor nutrition in early life produces permanent changes in glucose-insulin metabolism (Fig. 9.3).

Fetal malnutrition leads to impaired fetal growth and low birth weight favors a thrifty phenotype that is epigenetically programed to use nutritional energy efficiently, i.e. to be prepared for a future environment with low resources during adult life. In general, the phenotype of an individual is based on complex genome-environment interactions resulting in life-long remodeling of the epigenome. For example, even six decades after birth the individuals of the *Dutch Hunger Winter Families Study* showed lower DNA methylation levels at the regulatory region of the imprinted gene *IGF2* (Sect. 4.4). This epigenetic mark is associated with an increased risk of obesity, dys-lipidemia and insulin resistance, when the respective individuals are exposed to an obesogenic environment. This suggests that also for humans there is a link between pre-natal nutrition and epigenetic changes as described for rodents. Thus, environmental exposures, in particular during early life, can be stored as epigenetic memory.

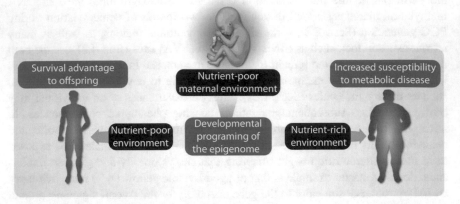

Fig. 9.3 Thrifty phenotype/DOHaD hypothesis. Intra-uterine stressors, including maternal undernutrition or placental dys-function (leading to impaired blood flow with consecutively hypoxia or reduced nutrient transport) can initiate abnormal patterns of development, histone modifications and DNA methylation. Additional post-natal environmental factors, including accelerated post-natal growth, obesity, inactivity and aging further contribute to the risk for T2D, potentially via changes in histone modifications and DNA methylation patterns of metabolic tissues (Sect. 13.4). Obviously, epigenetic changes during embryogenesis have a much greater impact on the overall epigenetic status of an individual than that of adult stem cells or somatic cells, since they affect far more following cell divisions

9.2   Population Epigenomics

The field of epigenetic epidemiology combines epigenomic methods (Sect. 2.2) with population-based epidemiological approaches. Different types of cohorts (Box 9.1) are studied with the goal to identify both the causes and phenotypic consequences of epigenomic variations. This implies that variations in epigenetic marks are assessed among individuals and throughout their lifetimes in response to their environmental exposures and lifestyle.

> **Box 9.1 Epigenomics of Populations**
> Different types of population-based study designs can provide samples for epigenomic analysis. The timing of studies is visualized in Fig. 9.4.
>   *Natural "experiments"*: Studies in which the exposure to a condition is not under experimental control. For example, in the Netherlands the *Dutch Hunger Winter Families Study* (www.hongerwinter.nl) follows individuals who were exposed as an embryo to famine of their mothers during the war winter of 1944/45.
>   *Longitudinal birth cohorts*: These studies analyze epigenetic changes over time and relate them to environmental exposures and the development of disease. For example, in the UK the *Avon Longitudinal Study of Parents*

(continued)

Box 9.1 Epigenomics of Populations (continued)
and Children (www.bristol.ac.uk/alspac) follows the health of 14,500 families, who had children born in 1991/92.

Longitudinal twin studies: Since mono-zygotic twins genetically identical, they are cross-wise the ideal references for the study of individual epigenetic variations. For example, in Australia the *Peri/Post-natal Epigenetic Twins Study* (www.mcri.edu.au/research/projects/peripostnatal-epigenetic-twins-study-pets-0) explores epigenetic variation between twins from birth on.

Pre-natal cohorts: The pre-natal developmental period is the most crucial time frame for epigenetic programing (Sect. 8.1). When individuals are tracked from the point of conception onwards, maternal and intra-uterine influences, their genotype and post-natal environment can be related to health and disease. For example, in the UK the *Southampton Women's Survey* (www.mrc.soton.ac.uk/sws) follows 12,500 women and their off-spring through their pregnancies.

In vitro fertilization (IVF) conception cohorts: The *Danish National IVF Cohort Study* studied frequency of imprinting diseases in children born after IVF.

Epigenetic changes can occur at any time during life, although increased sensitivity exist during early embryogenesis (Sect. 8.1, Fig. 9.4, *top*). Moreover, the individual's epigenome changes due to (1) environmental exposures and (2) stochastic epigenetic drift associated with aging (Sect. 9.3). Family studies (Box 9.1) are well suited to investigate epigenetic changes in the offspring based on environmental exposures of parents during gametogenesis. IVF cohorts, archived Guthrie cards and birth cohorts that track life from as early as periconception (i.e. around the time of conception) allow the study of epigenetic changes based on pre-natal environment and their association with disease phenotypes early in life (Fig. 9.4, *bottom*). Cohorts based on natural experiments, i.e. when the exposure to severe environmental conditions was not under experimental control, such as the *Dutch Hunger Winter* (Sect. 9.1), enable the investigation of the link of environmental exposures in early life with the onset of disease phenotypes decades later.

Prospective cohorts, in particular those involving mono-zygotic twins having identical genomes, parents, birth date and gender, in a longitudinally way study the contribution of age-related epigenetic modifications in common diseases, such as diabetes, autoimmune diseases (Chap. 12), cancer, Alzheimer disease and aut-ism spectrum disorders (Chap. 11). Interestingly, these studies have uncovered significantly increasing epigenetic variations between twins across their lifespan. Short-term interventions, such as dietary studies, can identify specific environmental exposures that lead to tissue-specific epigenetic changes. Finally, the follow-up of long-lived families helps to identify the impact of epigenetics for healthy aging (Sect. 9.3).

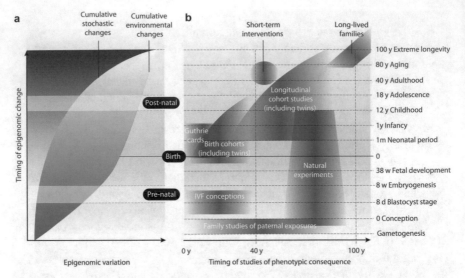

Fig. 9.4 Epigenetic variation in populations. Epigenetic changes can occur at any time during life, but there is significantly increased sensitivity during early pre-natal development (a). Pre-natal epigenetic changes may be investigated using IVF cohorts (Box 9.1), archived Guthrie cards and birth cohorts tracking life from as early as periconception (b). Historical famines represent the few opportunities to link the pre-natal environment to health outcomes later in life. Longitudinal cohort studies (especially involving twins) sample peripheral tissues and take biopsies from disease-relevant tissues. Short-term (dietary) interventions can identify specific dietary compounds that induce tissue-specific epigenetic modifications, while long-lived families can help in identifying the importance of maintaining epigenetic control for healthy aging (Sect. 9.3)

Most ongoing long-term studies were initially not designed for epigenetic analysis, i.e. they cannot incorporate all types of epigenomic assays needed for the validation of epigenetic associations. This led to the design of EWASs that measure the epigenetic susceptibility for a trait, such as for the phenotypic property height or the risk to develop T2D. For example, for hundreds to thousands of individuals statistically significant associations are measured between changes in DNA methylation at specific genomic loci and their impact on the expression of nearby genes possibly influencing disease risk. Like in a GWAS, the typical design of an EWAS is cross-sectional, i.e. it collects samples from individuals with a disease ("cases") and compares their epigenome with that of healthy individuals ("controls," Fig. 9.5). However, for EWASs a cross-sectional design is mostly not the best option, since it bears the possibility of reverse causation, in which the epigenome is influenced by the disease rather than causing it. For example, DNA methylation patterns in peripheral blood mononuclear cells (PBMCs) were found to change in response to increased BMI or altered blood lipid profiles. Therefore, the analysis of confounding variables, such as age, gender, smoking and cellular heterogeneity, and the collection of longitudinal data are essential, in order to correctly assess causality and to determine the biological relevance of disease-associated epigenetic variations.

Most EWASs focus on DNA methylation, since respective assays have for any given CpG a bimodal outcome of either being methylated or not. In contrast,

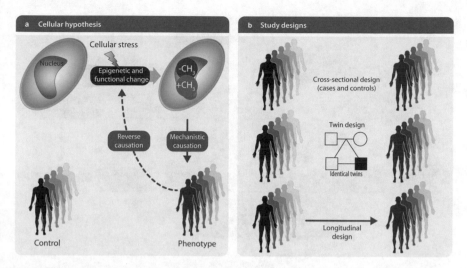

Fig. 9.5 Designing and interpreting EWAS. The cellular hypothesis of most EWASs is that epigenetic changes, such as different levels of DNA methylation, at specific genomic regions, affect gene expression at nearby genes (a). This then causes a phenotype (*purple*) in comparison to controls (*black*) and may provide a mechanistic explanation. It is also possible that the phenotype induces the epigenetic change, which is referred to as reverse causation. EWASs mostly have one of the three illustrated designs: (1) cross-sectional analysis comparing individuals with and without a phenotype, (2) studies of mono-zygotic twins discordant for a phenotype and (3) longitudinal studies of individuals before and after developing a phenotype (b)

ChIP-based assays are far less quantitative and collection of appropriate samples is more challenging. Some EWASs assess differential DNA methylation at individual CpGs independently from each other, such as in GWASs individual SNPs are analyzed. However, in many cases the analysis of the combined methylation profile of a genomic region, referred to as differentially methylated region (DMR), is more appropriate. Moreover, since there are both stochastic and age-related interindividual variations in the methylation status of DMRs (Sect. 9.3), it is more relevant to analyze in parallel several DMRs close to functionally related genes. The conclusions of an EWAS may be strongly suggestive of being causal and may provide ideas on the underlying mechanisms, when combined with additional data, but in its core every EWAS is still based on correlation analysis.

It is assumed that epigenetic changes only "fine-tune" biologically processes, i.e. in contrast to GWAS of mono-genetic diseases, where individual SNPs can have a large impact (i.e. a high "odds ratio") on the investigated trait, it is not expected that any EWAS will result in the identification of a DMR with a very large odds ratio. The functionally most relevant epigenomic variations occur at gene regulatory elements, such as at CpGs or transcription factor binding sites in promoter and enhancer regions. Interestingly, the DNA methylation levels at thousands of CpGs are affected by genetic variants, such as SNPs. These genomic sites are referred to as methylation quantitative trait loci (QTLs). The vast majority of epigenomic variations are in *cis* to their consequences on chromatin architecture and gene expression, i.e. they happen with the same genomic region, such as a

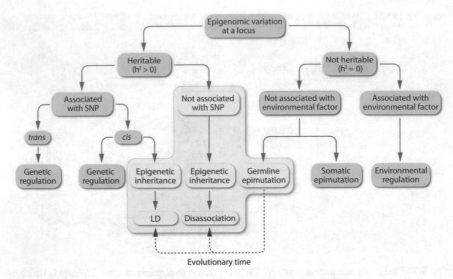

Fig. 9.6 Epigenomic variations. The epigenomic variation of a genomic region can be considered as a QTL. A number of different methods, such as variance components analysis on the basis of pedigree data, can be applied, in order to estimate epigenetic heritability (h^2). In cases of epigenetic inheritance ($h^2 > 0$) the epigenomic variant can be associated with a SNP and involved in gene regulation, which is either in *trans* or in *cis*. The genetic architectures of the latter can be detected by various methods including EWASs. When there is no inheritance ($h^2 = 0$), the epigenomic variation could be associated with environmental exposures in the past or presence. In the absence of causative environmental factors, the epigenomic variation could be the result of an epimutation (Sect. 9.3) in somatic or germ cells that leads to an epigenetic drift (Fig. 9.2). Epimutations are preferentially detected by single-cell methods (Box 2.1)

TAD. In contrast, *trans*-acting epigenomic variations, like epimutations in pleiotropic regulators, such as the master transcription factor OCT4, are very sparse suggesting that they may be highly deleterious. In fact, they are one basis of the epigenomics of cancer (Sect. 10.1).

In general, epigenomic variations have different properties, such as being (1) heritable or not, (2) associated with a SNP or not, associated with environmental factors or not, (3) in *trans* or *cis* concerning gene regulation or (4) an epimutation in somatic or germ cells (Fig. 9.6). Epigenomic variations are determined by EWASs in combination with variance components analysis, linkage mapping methods or single-cell epigenomic sequencing techniques (Box 2.1). In particular, the correlation of EWAS results with transcriptome data from the same individual is an important validation step, since differential gene expression is the primary readout of functionality. Moreover, cell type-specific epigenome reference data from ENCODE, Roadmap Epigenomics and IHEC (Sect. 2.2) are of key importance for the integration of epigenome and transcriptome data.

Taken together, the epigenome of individuals is far more variable than their genome. Twin studies indicated that, for example, the heritability of DNA methylation is in the order of 18–37%. As expected, genome-wide surveys demonstrated that the patterns of accessible chromatin, DNA methylation and histone modifications are more similar among relatives, such as father-mother-child trios, than with unrelated individuals.

9.3 Epigenomics of Aging

Like in other animal species, also in humans lifespan comprises (1) a period of growth and differentiation that ends up in sexual maturity, i.e. in a period of maximal fitness and fertility, and (2) a period of aging that comes with loss of function at the various levels of cells, tissues and the organism as a whole. The end of the latter phase is associated with a wide range of diseases, including metabolic disorders, cardiovascular and neurodegenerative diseases, and many cancers. Accordingly, healthspan is defined as the period of disease free health, represented by high cognition and mobility. Thus, understanding the changes that occur during aging, i.e. the hallmarks of aging, and identifying regulators of lifespan and healthspan is a key question for everyone.

Individuals have a personal rate of aging that depends on gender – women tend to live longer than men – lifestyle choices, such as smoking or physical inactivity, environmental factors, and the more. There is a clear genetic basis for longevity, but various animal aging models agree on that also non-genetic factors, such as calorie restriction, low basal metabolic rate, increased stress response and reduced fertility, may play a major role in determining the lifespan of individuals. The molecular basis of all non-genetic factors are cellular perturbations that result in the modulation of signal transduction pathways affecting the epigenome, i.e. epigenetic changes are a major contributor to the aging process. Thus, not only diseases (Sect. 9.1), but also aging result in epigenetic drifts. In humans the non-genetic contribution to aging was estimated to be some 70%. Changes in the epigenome also can occur spontaneous or "stochastically" without the contribution of a cellular perturbation. Nevertheless, the likelihood for stochastic changes of the epigenome is increased by chemicals disrupting DNA methylation marks or by errors in copying the methylation status during DNA replication.

There are molecular markers of age, such as telomere length or the expression of genes in metabolic and DNA repair pathways, that are sensitive to environmental stress. Moreover, changes of the epigenome, in particular in DNA methylation (Fig. 9.7), are associated with the chronological age of an individual as well as with age-associated diseases, such as the metabolic syndrome and cancer. Typically, the DNA methylation status at key CpGs is measured from accessible tissues and cell types, such as skin or liver biopsies and PBMCs. The respective chromatin landscapes are then correlated with chronological age and biological age, i.e. the age at which the population average is most similar to the individual.

In general, DNA methylome profiling of a large cohort of individuals spanning over a wide age range provide a good correlation between chronological and biological age, but there are also significant inter-individual variations. At a given chronological age the investigated tissue of some individuals has a far "younger" epigenome, while that of others is already "older" (Fig. 9.7). Accordingly, it can

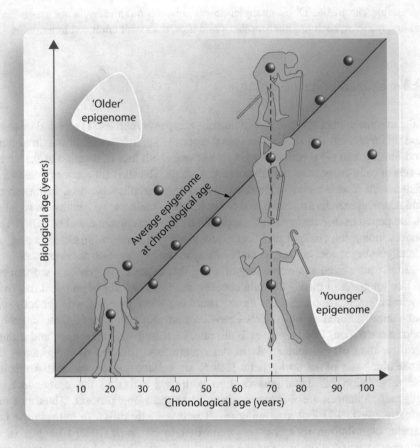

Fig. 9.7 Epigenetic biomarkers of age. Epigenomic patterns not only monitor cellular identities, but also cellular health and age. For example, changes in the DNA methylation status of CpGs, as measured in PBMCs taken from individuals of different age (*symbols*), can serve as a sensor for chronological age. However, there is significant deviation from the linear fit (*diagonal line*), which suggests that methylation patterns also represent biological age

be expected that the latter individuals may have an earlier onset of age-related diseases from the respective tissue and eventually may die in younger years that than the former individuals. This has been observed with individuals suffering from progeroid syndromes, such as Down syndrome. In contrast, the blood of the offspring of super-centenarians, i.e. individuals who reached an age of at least 105 years, has a lower epigenetic age than that of age-matched controls.

Moreover, the biological age of a specific tissue, for example, that of the liver of an obese person, may be significantly older than other tissues, such as blood or muscle, from the same individual. DNA methylation-based age of, for example, PBMCs, which can be considered as an epigenetic clock, is a more accurate predictor for mortality by all causes in later life than other biomarkers of aging, such as changes in telomere length. Interestingly, ES and iPS cells show in these assays

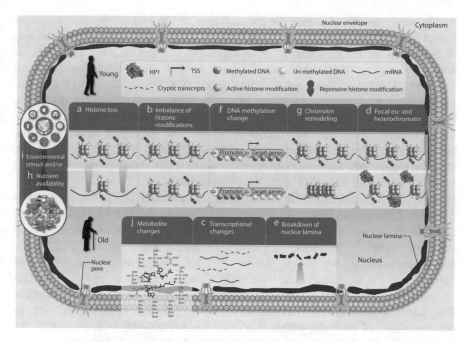

Fig. 9.8 Epigenetics of senescence and aging. The epigenetic hallmarks of senescence and aging are a loss of histones (a), imbalance of activating and repressive modifications (b), changes in gene expression (c), losses and gains in heterochromatin (d), breakdown of nuclear lamina (e), global hypo-methylation and focal hyper-methylation (f) and chromatin remodeling (g). These changes are heavily dictated by environmental stimuli (h) and nutrient availability (i) that in turn alter intra-cellular metabolite concentrations (j)

as ageless. Similarly, sperm cells are estimated younger than somatic cells from the same individual. Interestingly, epigenetic clocks of mice (average lifespan some 2 years) tick faster than those of humans (average lifespan some 80 years). Moreover, a quantitative model of the aging methylome is able to distinguish relevant factors in aging, including gender and genetic variants.

In addition to changes in DNA methylation the key epigenetic hallmarks of aging (Fig. 9.8) include (1) a general loss of histones due to local and global chromatin remodeling, (2) an imbalance of activating and repressive histone modifications, (3) site-specific loss and gain in heterochromatin, (4) significant nuclear reorganization and (5) transcriptional changes.

The general loss of histones in aging cells is tightly linked to cell division (Fig. 9.8a). For example, nuclear blebs of senescent human cells contain a large number of histones. Senescence is a cell cycle arrest in response to stress. *In vivo* this is linked to age-associated tissue decline. Human cells then develop senescence-associated heterochromatin foci. These are regions of highly condensed chromatin associated with heterochromatic histone modifications, heterochromatic proteins and histone variant macroH2A (Sect. 3.3). In general, in senescent cells on some 30% of the genome, chromatin is reorganized with an increase of H3K4me3 and H3K27me3 marks within LADs and a loss of H3K27me3 outside of LADs (Fig. 9.8b).

The most significant molecular consequence of the loss of repressive histone marks and the gain of activating marks during aging is a change in gene expression, such as the up-regulation of genes related to cell proliferation, cell adhesion and ribosomal proteins, while genes related to cell cycle, DNA base excision repair and DNA replication are down-regulated (Fig. 9.8c). The lack of some of the repressed genes negatively affects longevity, while some of the activated genes are detrimental to lifespan. Moreover, the transcriptional reprograming happening during onset of senescence also occurs through altered activity of chromatin modifiers and remodelers. In addition, aging is coupled with a stochastic de-regulation of gene expression, referred to as transcriptional noise. Interestingly, older cells tend to have higher transcript levels than younger cells.

Constitutive heterochromatin at telomeres, centromeres and pericentromeres is established during embryogenesis (Sect. 8.1) and is thought to be maintained throughout lifespan. However, senescent cells loose some of these regions of constitutive, chromatin resulting in growth of euchromatic regions (Fig. 9.8d). Moreover, a loss of the nuclear lamina (Fig. 9.8e) stimulates the breakdown of heterochromatin organization and the re-localization of heterochromatic proteins to regions in the genome where they contribute to the formation of region-specific foci. Laminopathies display deleterious changes to nuclear organization, such as the expression of progerin, a truncated dominant-negative lamin A protein, in the Hutchinson-Gilford progeria syndrome (HGPS). Cells from HGPS patients show abnormalities in nuclear morphology, defect DNA damage repair, changes in chromosome organization, increased rates of cellular senescence and many alterations in heterochromatin proteins, such as low levels of HP1, H3K9me3, and H3K27me3 and increased levels of H4K20me3. Moreover, the deficiency in an ATP-dependent helicase RECQ3 causes the Werner syndrome and leads to global loss of chromatin compaction, decreased H3K9me3 and H3K27me3 levels and increased phosphorylation of the histone variant H2A.X at centromeres.

The already discussed global hypo-methylation and local hyper-methylation of genomic DNA during aging fits with the observation of global heterochromatin de-regulation in combination with focal increase at some genomic regions (Fig. 9.8f). DNA methylation is primarily lost at repetitive genomic regions that are in constitutive heterochromatin, while hyper-methylation mostly occurs at CpGs close to TSS regions. SWI/SNF chromatin remodelers are associated with gene activation and they seem to promote aging, while repressive ISWI chromatin remodelers support longevity (Fig. 9.8g, Sect. 7.1). The nucleosome remodeling and deacetylase (NuRD) remodeling complex is down-regulated in cells obtained from HGPS patients as well as in older healthy donors. The consequence of NuRD down-regulation is a loss of senescence heterochromatic features.

Dietary restriction extends the lifespan in many model organisms, such as yeast, worms, flies and even primates. Gene expression changes related to dietary restriction, such as up-regulation of SIRT genes, promote global preservation of genome integrity and chromatin structure, such as maintenance of heterochromatin (Fig. 9.9). Evolutionary highly conserved proteins, such as insulin and IGF receptors, the amino acid sensor target of rapamycin (TOR) and the NAD^+-sensing

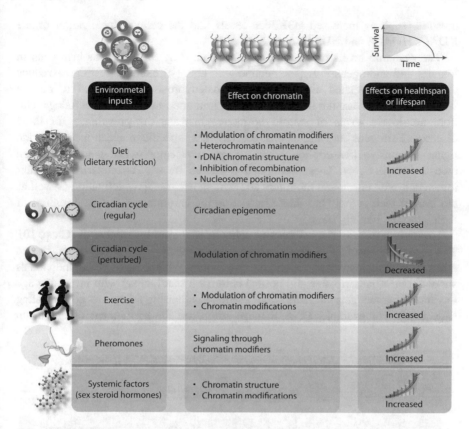

Fig. 9.9 Effects of environmental inputs on longevity and chromatin. Many environmental signals that modulate lifespan also affect chromatin. These are dietary restriction, the circadian cycle, physical activity and sex steroid hormones. More details are provided in the text

KDAC SIRT1, integrate metabolic signals into chromatin responses. They inform the epigenome on nutrient availability (Chap. 13) and thus have a key role in determining lifespan (Fig. 9.8h). Similar principles apply to the sensors of other environmental inputs (Fig. 9.8i) or intra-cellular metabolites (Fig. 9.8j), such as steroid hormones via their nuclear receptors or α-KG via TET enzymes, that affect longevity via changes in the chromatin landscape.

The circadian clock is based on the rhythmic recruitment to chromatin of the transcription factors CLOCK, PER2 and ARNTL that control a number of physiological functions in a 24 hours cycle. CLOCK is a KAT with specificity for H3K9 and H3K14. In contrast, the NAD^+ sensor SIRT1 deacetylates PER2 and ARNTL. Disrupting the circadian clock negatively influences health and longevity, which restoring a functional circadian clock in aging animals improves health and lifespan (Fig. 9.9). Physical activity promotes healthy aging, as it prevents cognitive decline and is associated with a 30% reduction in all-cause mortality. Interestingly, physical activity induces changes in the chromatin of skeletal muscles, such as increased H3K36ac levels and the cellular localization of the KDACs HDAC4 and HDAC5.

The systemic levels of sex steroid hormones (Fig. 9.9), such as estrogens in females and androgens in males, decline with age. For example, estrogens reduce the risk for age-related diseases, such as osteoporosis, sarcopenia (i.e. muscle weakness), cardiovascular diseases, reduced immune function and neurodegeneration, i.e. lower levels of steroid hormones increase the prevalence for these diseases. Estrogens and androgens act via their specific nuclear receptors, the ligand-responsive transcription factors estrogen receptor α and β and androgen receptor, respectively. These transcription factors are well characterized for their interaction with chromatin modifiers and remodelers, i.e. their activation as well as the lack of their activity has direct impact on the chromatin at the local regions of their genomic binding sites.

Genomic instability is a hallmark of aging as well as of tumorigenesis (Chap. 10). The accumulation of DNA mutations or even aneuploidy, as during aging, has an effect on both the transcriptome and the epigenome. While cells of young individuals show a robust transcriptome and normal chromatin states, along with increasing age the transcriptome gets unstable, and aberrant chromatin states are accumulating (Fig. 9.10). For example, DNA damage stimulates the recruitment of chromatin

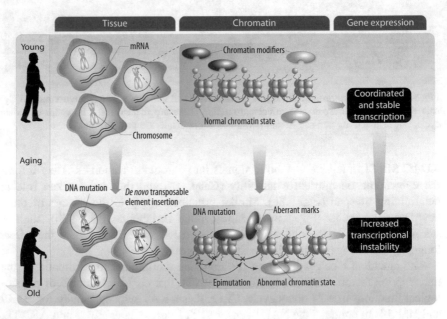

Fig. 9.10 Epigenomic and transcriptomic changes link with genomic instability during aging. Cells of young individuals (*top left*) have robust transcriptomes (shown as consistent mRNA levels), genomic integrity is kept (shown as intact chromosomes that have either a low mutation rate or are repaired) and the epigenome/chromatin stays normal. In contrast, during aging (*bottom left*) the transcriptome and the genome both get unstable due to the accumulation of DNA mutations and epimutations resulting in aberrant chromatin marks and states

modifiers that may induce abnormal chromatin states. In turn, epigenomic changes during aging can increase the susceptibility of the genome to mutation and in parallel reduce the precision of transcription. Moreover, errors in DNA repair and failure to correctly replicate the genome and epigenome not only increase the number of DNA mutations but also of epimutations (Sect. 9.2). Thus, since genome surveillance and epigenetic remodeling influence each other, environment-induced epigenomic instability throughout life is an important driver of the aging process.

Key Concepts
- The epigenome has memory functions, since it is able to preserve the results of cellular perturbations by environmental factors in form of changes in DNA methylation, histone modifications or 3D organization of chromatin.
- Although different persons show consistency in the overall epigenome patterns of their tissues, individuals vary far more on the level of their epigenomes than on the level of their genomes.
- The epigenome is able to memorize lifestyle events in basically every tissue or cell type, i.e. the memory of an individual is not only stored by neurons, but also the immune system memorizes encounters, for example with microorganisms, and metabolic organs remember personal habits on diet and physical activity, respectively.
- Short-term "day-to-day" responses of the epigenome are primarily mediated by non-inherited changes in the histone acetylation level, while long-term decisions, for example, concerning cellular differentiation, are stored in form of histone methylation marks. In addition, DNA methylation is particularly suited for a long-term memory of cells.
- Epigenetic changes can occur at any time during life, although increased sensitivity exist during early embryogenesis.
- Epigenetic drifts can be detected in epigenomic patterns, such as DNA methylation maps that are heritable from parent to daughter cells and may in part even be transferred to the next generation.
- The concept of transgenerational epigenetic inheritance could explain how lifestyle factors of parents and grandparents, such as daily diet and physical activity, can affect their offspring.
- The thrifty phenotype hypothesis suggests that poor nutrition in early life produces permanent changes in glucose-insulin metabolism.
- EWASs use genome-wide assays, in order to demonstrate that *cis*- and *trans*-regulatory mechanisms shape patterns of population epigenomic variations in detail and as a whole.
- The typical cross-sectional design of EWASs bears the possibility of reverse causation, in which the epigenome is influenced by the disease rather than causing it.
- Different types of human cohort studies, ideally composed of monozygotic twins and best having a longitudinal design, are well suited to

(continued)

Key Concepts (continued)
  identify population epigenomic variations that are associated with human traits, such as disease risk.
- Epigenetic changes are a major contributor to the aging process, i.e. not only diseases, but also aging result in epigenetic drifts.
- Genome surveillance and epigenetic remodeling mutually influence each other, i.e. environment-induced epigenomic instability throughout life is an important driver of the aging process.
- Common hallmarks of aging are associated with specific chromatin patterns, and chromatin modifiers are able to modulate both life- and healthspan.
- Epigenomic signatures can serve as biomarkers of aging and may be druggable targets, in order to delay or reverse age-related disease.

Additional Reading

Benayoun BA, Pollina EA, Brunet A (2015) Epigenetic regulation of ageing: linking environmental inputs to genomic stability. Nat Rev Mol Cell Biol 16:593–610

Birney E, Smith GD, Greally JM (2016) Epigenome-wide association studies and the interpretation of disease-omics. PLoS Genet 12:e1006105

Heard E, Martienssen RA (2014) Transgenerational epigenetic inheritance: myths and mechanisms. Cell 157:95–109

Mill J, Heijmans BT (2013) From promises to practical strategies in epigenetic epidemiology. Nat Rev Genet 14:585–594

Sen P, Shah PP, Nativio R et al (2016) Epigenetic mechanisms of longevity and aging. Cell 166:822–839

Taudt A, Colome-Tatche M et al (2016) Genetic sources of population epigenomic variation. Nat Rev Genet 17:319–332

Chapter 10
Cancer Epigenomics

Abstract Large-scale cancer genomic projects indicated that more than 50% of human cancers carry mutations in key chromatin proteins. Moreover, compared to normal cells cancer cells show genome-wide changes in DNA methylation, histone modifications and 3D chromatin structure. In addition, many tumors re-activate programs of fetal development, which is a sign of epigenetic reprograming. The mechanistic bases of cancer epigenomics are specific genetic, environmental and metabolic stimuli that disrupt the homeostatic balance of chromatin, which then either becomes very restrictive or permissive. Restrictive chromatin states can prevent the induction of tumor suppressor programs or block differentiation. In contrast, permissive states allow oncogene activation or non-physiologic cell fate transitions.

Many of these epigenetic changes are only "passengers," but a few also act as "drivers" of the tumorigenesis process. In pre-malignant cells these epigenetic dys-regulations promote tumor initiation and in malignant cells they accelerate tumor evolution and adaptation. The effects of diverse oncogenic stimuli are mediated via epigenetic modulators and contribute to diverse aspects of cancer biology, such as all hallmarks of cancer. Recent drug discovery efforts targeted the epigenome and several new drugs – inhibitors of chromatin modifiers – are tested in clinical trials and some were already approved by the US Food and Drug Administration (FDA).

In this chapter, we will discuss the impact of epigenetic modifiers in the tumorigenesis process. We will present epigenetic dys-regulations as the basis for an epigenetic reprograming that can transform normal cells into tumor cells. In this context, we will understand how epigenomic changes will contribute to basically all hallmarks of cancer. Finally, we will analyze the impact of epigenetic therapy for the treatment of cancer.

Keywords Epigenetic modifiers · epigenetic reprograming · tumorigenesis · epigenetic mediators · hallmarks of cancer · epigenetic modulators · KMT inhibitors · KDM inhibitors · KDAC inhibitors · DNMT inhibitors · epigenetic therapy

© Springer Nature Singapore Pte Ltd. 2018   159
C. Carlberg, F. Molnár, *Human Epigenomics*, https://doi.org/10.1007/978-981-10-7614-5_10

10.1   Impact of Epigenetic Mutations in Cancer

Cancer is typically considered a disease of the genome, but it also comes along with abnormalities in gene expression, cellular identity and responsiveness to internal and external stimuli indicating that also changes in the epigenome play a profound role in tumorigenesis. In fact, most types of cancer carry mutations both in their genome and epigenome. Childhood tumors are often based only on a small number of genetic mutations, but these often occur in genes that encode for chromatin modifiers. For example, the biallelic loss of the *SMARCB1* gene, which encodes for a member of SWI/SNF remodeling complex, in pediatric rhabdoid tumors was a first indication that the disruption of epigenetic control can serve as a "driver" for cancer. In total, some 50% of human cancers harbor mutations in chromatin proteins. It is also important to realize that the epigenetic signature of a cell allows more variation than its genetic status, since the error rate, for example, in inheritance of DNA methylation is some 4% for a given CpG per cell division, while the mutation rate of the genome during DNA replication is far lower. Thus, epigenetic variability leads in much shorter time to phenotypic selection, such as possible onset of cancer, than genome mutations.

Already some 35 years ago, epigenetics was first associated with cancer, when it was found that (1) the genomic DNA of tumor cells was globally less methylated and (2) regulatory regions of tumor suppressor genes, such as *TP53*, were more methylated than non-malignant reference cells of the same individual (Fig. 10.1, *top*). The hyper-methylation of tumor suppressor genes is an example of an epigenetic drift (Sect. 9.1). It results in their transcriptional silencing and thus provides a reasonable mechanism for the increased growth potential of cancer cells. Like in aging (Sect. 9.3), genome-wide hypo-methylation also contributes to cancer, because it promotes genomic instability and inappropriate activation of oncogenes and transposons.

From the genetic perspective cancer genes can be classified into dominant oncogenes, which can be activated by gain-of-function mutations, amplifications or translocations, and recessive tumor suppressor genes, the expression of which is often lost, for example, by loss-of-function mutations or methylation of their promoter regions. An alternative classification divides them into "drivers," the mutation of which directly affects tumorigenesis, and "passengers," which are mutated as a side product, but do not have a functional contribution on oncogenesis. The epigenetic perspective adds a further classification of cancer genes as encoding for epigenetic modifiers, mediators and modulators. Epigenetic modifier genes encode for proteins that directly modify the epigenome through DNA methylation, histone modification or structural changes of chromatin. Epigenetic mediators are targets of epigenetic modification and largely overlap with the genes involved in the reprograming of somatic cells into iPS cells (Sect. 8.5). In signal transduction pathways, which are stimulated by environmental molecules, inflammation and other forms of stress, epigenetic modulators are located upstream of modifiers and mediators. The actions of epigenetic modulators are often the first steps in tumorigenesis resulting in changing epigenome patterns.

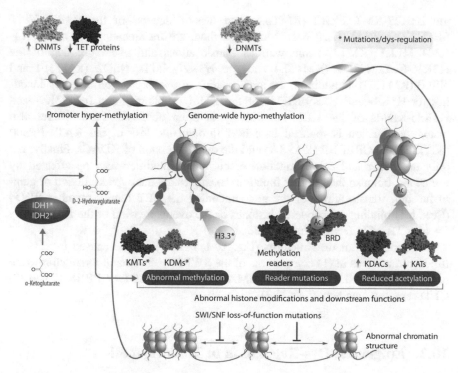

Fig. 10.1 Epigenetic mutations in cancer. There are four major types of epigenetic mutations affecting cancer: DNA hyper-methylation at promoters (*top left*), genome-wide DNA hypo-methylation (*top right*), abnormal modification of histones and/or their recognition (*center*) and abnormal chromatin structures caused by mal-functional chromatin remodelers (*bottom*). More details are provided in the text

Epigenetic mutations are either gain-of-function or loss-of-function. There are already a number of small molecule inhibitors for proteins affected by gain-of-function mutations, while targeting loss-of-function mutations remains difficult (Sect. 10.5). Loss-of-function mutations in genes encoding for DNA demethylases (*TET1*, *TET2* and *TET3*) or increased expression of genes for encoding DNMTs (*DNMT1*, *DNMT3A* and *DNMT3B*) can cause promoter hyper-methylation in some cancers. In contrast, genome-wide hypo-methylation is often based on loss-of-function mutations in the *DNMT3A* gene. Mutations in the genes encoding for the metabolic enzymes isocitrate dehydrogenase (IDH) 1 and IDH2 produce from α-KG the "oncometabolite" 2-hydroxyglutarate that inhibits TET enzymes and KDMs (Fig. 10.1, *center*). This leads to the increased methylation of both DNA and histones.

There are a number of both gain-of-function and loss-of-function mutations in genes encoding for chromatin modifiers. Abnormal histone methylation may be caused by mutations in genes encoding for KMTs and KDMs as well as for the histone variant H3.3 (reducing the genome-wide methylation of H3K27 and H3K36, Fig. 10.1, *center*). Examples are gain-of-function and overexpression of the H3K27-KMT EZH2 (KMT6A) and loss-of-function of the H3K36-KMT SETD2 (KMT3A) (Sect. 6.2). Moreover, there are translocations of the H3K4-KMT MLL1 (KMT2A) and well as translocations and overexpression of the H3K36-KMT NSD1 (KMT3B) and the H3K27-KMTs NSD2 (KMT3F) and NSD3 (KMT3G). In addition, the amplification or overexpression of genes encoding for H3K4- and H3K9-KDMs of the KDM1 subfamily and for H3K9- and H3K36-KDMs of the KDM4 subfamily is often observed. Furthermore, also histone acetylation is reduced in cancer through the loss of the KATs EP300 (KAT3B) and CREBBP (KAT3A) and the overexpression of KDACs. Finally, not only the writer and eraser function of chromatin modifiers can be affected by mutations, but also their reader function. Examples are the overexpression or gain-of-function translocations of the gene encoding the BET family member BRD4 (Sect. 6.1) binding to acetylated histones or the overexpression of the PHD family member TRIM24 recognizing H3K23ac.

Abnormal chromatin structures (Fig. 10.1, *bottom*) can be caused by loss-of-function mutations in (1) components of the SWI/SNF chromatin remodeling complex, such as ARID1A, ARID1B, SMARCA4, SMARCB1 or PBRM1, and (2) CHD complex members (Sect. 7.2).

10.2   Epigenetic Dys-Regulation in Tumorigenesis

Changes in DNA methylation patterns are key epigenetic dys-regulations occurring during tumorigenesis. Compared with normal cells of the same individual, the epigenome of tumor cells shows a massive overall loss of DNA methylation, while for certain genes also hyper-methylation at CpGs can be observed (Fig. 10.2). This so-called CpG island methylator phenotype (CIMP) is the best-known epigenetic dys-regulation in cancer.

Global DNA hypo-methylation during tumorigenesis (1) generates chromosomal instability, (2) reactivates transposons and (3) causes loss of imprinting. Low DNA methylation favors mitotic recombination leading to deletions and promotes chromosomal rearrangements, such as translocations. The disruption of genomic imprinting, such as the loss of imprinting of the *IGF2* gene (Sect. 4.4), is a risk factor for different types of cancer, such as colon cancer or Wilms tumor.

Hyper-methylated promoter regions of tumor suppressor genes, such as *TP53*, *RB1* and *MGMT*, can serve as biomarkers that provide significant diagnostic potential in the clinic, in particular in early-detection screenings of individuals with a high familial risk of developing cancer. Many CpGs can become methylated already early in tumorigenesis. In majority they rather affect the expression of genes involved in carcinogen metabolism, cell-to-cell interactions and angiogenesis than classical tumor suppressor genes that control the cell cycle, DNA repair and apoptosis. The profiles of CpG hyper-methylation vary with tumor types. Each type of cancer can be characterized by its specific DNA hyper-methylome, i.e. these epigenetic marks are comparable to traditional genetic and

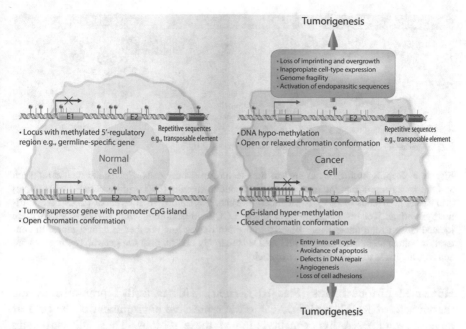

Fig. 10.2 Changes of DNA methylation patterns during tumorigenesis. Compared to normal cells (*left*) cancer cells are hypo-methylated on the genome-wide scale (*top right*), in particular at repetitive sequences, such as transposons. In addition, imprinted and tissue-specific genes often get demethylated. Hypo-methylation causes changes in the epigenetic landscape, such as the loss of imprinting, and increases the genomic instability that characterizes cancer cells. Another common alteration in cancer cells is the hyper-methylation of CpGs within regulatory regions of tumor suppressor genes (*bottom right*). These genes are then transcriptionally silenced, so that cancer cells lack functions, such as inhibition of the cell cycle

cytogenetic markers. From about 200 genes that are regularly mutated in various forms of human breast and colon cancers in average 11 carry a mutation in a single tumor type. For comparison, 100–400 CpGs close to TSS regions are found to be hyper-methylated in a given tumor, i.e. epigenetics is able to provide 10-times more information than genetics.

The changes in DNA methylation during tumorigenesis are always combined with other epigenetic dys-regulations, such as aberrant patterns of histone modifications and overall changes in the nuclear architecture, i.e. the overall epigenetic landscape of cancer cells is significantly distorted compared to somatic stem cells or differentiated cells (Fig. 10.3). These alterations in the 3D organization of chromatin exemplify epigenome changes during tumorigenesis. In differentiated cells developmentally repressed genes, i.e. genes that are not needed in a given cell type, are often found within LADs that constitutively localize close to the nuclear periphery (Sect. 3.5). A significant fraction of these LADs overlaps with so-called large organized chromatin K9-modifications (LOCKs), which are genomic regions that are enriched in repressive H3K9me2 and

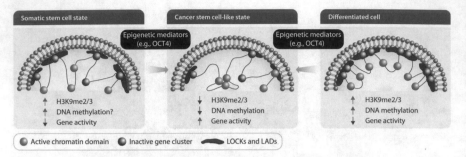

Fig. 10.3 Reprograming of the nuclear architecture in cancer cells. Epigenetic mediators, such as OCT4, can reprogram the epigenome of somatic stem cells (*left*) or differentiated cells (*right*) into cancer stem cells (*center*). Normal cells are characterized by high levels of H3K9 di- and tri-methylation as well as DNA methylation in LOCKs overlapping with LADs. The latter are located close to the nuclear membrane and contain only a low number of active genes. In contrast, in cancer stem cells LOCKs and LADs are largely absent, and a larger variety of genes are active. This leads to phenotypic heterogeneity

H3K9me3 histone marks (Fig. 10.3, *right*). This is further promoted by the recruitment of KDMs and KDACs to the repressive environment of the nuclear envelope and DNA hyper-methylation at these regions. Thus, 3D chromatin compaction mediates gene repression during lineage specification and represents a form of epigenetic memory. This results in reduced transcriptional noise and provides barriers for dedifferentiation. Although adult stem cells are in a less differentiated state than terminally differentiated cells, also in them specific LAD/LOCK structures are found (Fig. 10.3, *right*).

Proteins that regulate the interaction of chromatin with the lamina and recruit chromatin modifiers to the nuclear periphery function as epigenetic mediators. For example, re-activated pluripotency transcription factor OCT4 (Sect. 8.3) can reprogram the epigenome of both differentiated cells and adult stem cells into cancer stem cells (Fig. 10.3, *center*). The activation of the epigenetic mediator dissolves most of the LADs/LOCK structures, as a consequence of which a number of genes are re-activated. This provides cancer cells with phenotypic heterogeneity, such as increased variability in gene expression, in order to switch between different cellular states within the tumor. The loss of LOCKs also affects enhancer-TSS region communication within and between TADs, so that oncogenic super-enhancers are able to cluster. A similar process happens during epithelial-to-mesenchymal transition (EMT), which is a key process in normal wound healing, but also the first step towards metastasis. In EMT the activation of an H3K9-KDM, such as LSD1 (KDM1A), often is the initiating epigenetic event. When cancer cells have destabilized the epigenetic memory of the cells they originate from and form EMT-related chromatin structures, they gain phenotypic plasticity. Thus, the overall result of the change in chromatin architecture is an oncogenic transformation of the cell.

Different cancer types show a high phenotypic variability, which is largely based on their individual history of epigenetic dys-regulation. Based on Waddington's model of an epigenetic landscape (Sect. 1.4), the epigenetic status

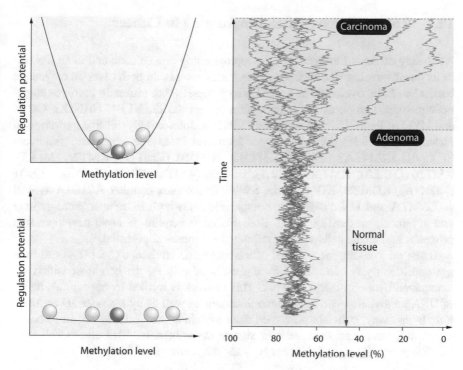

Fig. 10.4 Model of epigenetic dys-regulation. DNA methylation is used here as an example for epigenetic dys-regulation. The methylation level of a normal cell is illustrated as a ball at the bottom of a valley (*top left*), where regulatory forces allow only minor changes in the epigenetic status. In contrast, during tumorigenesis (*bottom left*) the landscape flattens, and the methylation levels can be far more variable. When the methylation level is modeled over time for 10 examples (*right*), the variations in the transition from normal tissue to adenoma and carcinoma become obvious as wider ranges

of a cell, such as its methylation level (Fig. 10.4), can be represented by a ball trapped in a valley. In case of normal differentiated cells, the borders of the valley are high and gene regulatory networks keep the cells in stable epigenetic homeostasis. This prevents the epigenetic state from moving too far from its equilibrium point in normal tissue (Fig. 10.4, *top left*). In contrast, a dys-regulation of the epigenome during tumorigenesis, such as overexpression of an epigenetic modulator (Fig. 10.3) or an inflammatory insult (Chap. 12), flattens the valley (Fig. 10.4, *bottom left*). Under these conditions of reduced regulation, the epigenetic status is more relaxed and influenced by stochastic variations. Thus, during tumorigenesis, DNA methylation levels diffuse away from the initial state in normal cells (Fig. 10.4, *right*). A plot of CpG methylation levels during the transformation of normal cells into adenoma and carcinoma cells shows a tight distribution in normal tissue, but a progression from adenoma to carcinoma. This explains the substantial level of epigenetic variation for a given cancer type across individuals or between metastatic cells originating from the same primary tumor. Accordingly, there is no defined epigenetic signature for cancer.

10.3   Role of Epigenetic Reprograming in Cancer

As already discussed in Sect. 8.5, the reprograming of a somatic cell to an iPS cell or its transformation to a cancer cell are related events. In both cases an epigenetic barrier has to be overcome in a multi-step processes that primarily involves epigenetic mediators, such as the transcription factors SOX2, KLF4, NANOG, OCT4 and MYC or the RNA-binding protein LIN28A. Interestingly, all five transcription factors are encoded by oncogenes. Moreover, also the chromatin modifiers SUV39H1 (KMT1A), EHMT2 (KMT1C), SETDB1 (KMT1E), KMT2A (MLL1), KMT2D (MLL2), KMT2C (MLL3), DOT1L (KMT4), EZH2 (KMT6A), LSD1 (KDM1A), KDM2B, KDM6A, the SWI/SNF complex member ARID1A as well as DNMTA and DNMT3B have comparable roles both in cellular reprograming and in tumorigenesis (Fig. 10.5). Both processes acquire de novo developmental programs and create cells with an unlimited self-renewal potential.

Based on Waddington's model of the epigenetic landscape (Sect. 1.4) cell lineage and identity is constrained by the walls of valleys, the height of which are determined by a chromatin network. This network is formed by appropriate levels of DNA methylation and histone modifications as well as by a proper 3D architecture. In this way, cells are prevented from switching states. However, in response to relevant intra- and extra-cellular signals the epigenome also allows cell state transitions, as illustrated in energy state diagrams (Fig. 10.6, *top left*). When

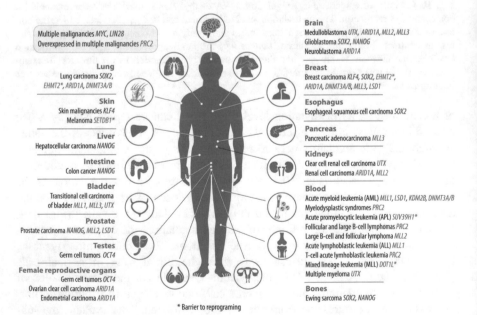

Fig. 10.5 Overlap of iPS nuclear reprograming and cancer. The same transcription factors (*red*) and chromatin modifiers (*green*) play a central role both in iPS cellular reprograming and in different types of cancer. Some of them are encoded by oncogenes and tumor suppressor genes

chromatin homeostasis is disturbed, for example, by epimutations, the cells do not respond appropriately to these signals. Overly restrictive chromatin networks create epigenetic barriers that prevent all types of cell state transitions (Fig. 10.6, *center*). In contrast, excessively permissive chromatin networks have very low barriers and allow multiple types of cell state transitions (Fig. 10.6, *bottom*). In brief, deviations from the norm are a major factor in tumorigenesis.

On the mechanistic level (Fig. 10.6, *center column*) the three scenarios can be explained by the actions of a KMT for repressive H3K27me3 marks, such as EZH2 (KMT6A), and a KMT for activating H3K4me3 marks, such as MLL1 (KMT2A). In normal cells both KMTs and their histone marks are in balance resulting in bivalent, poised constitutive heterochromatin at TSS regions (Sect. 5.2). The respective genes are transcribed only in response to appropriate stimuli. In restricted cells, EZH2 may have a gain-of-function epimutation, such as often observed in several forms of lymphoma, resulting in far higher levels of repressive H3K27me3 marks, stable heterochromatin and no gene transcription. EZH2 is the catalytic core of repressive PRC2 complexes and plays a major role in B cell development (Chap. 12).

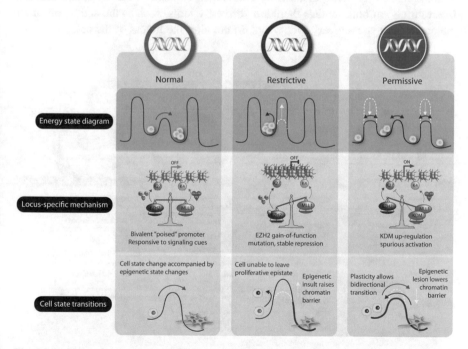

Fig. 10.6 Chromatin structure, cellular identity and cell state transitions. In normal cells (*left*) networks of chromatin proteins stabilize the states of cells but also mediate the response to intra- and extra-cellular stimuli and occasionally allow cell state transitions. However, cells in which the chromatin network is perturbed do not respond appropriately. In restrictive chromatin (*center*) epigenetic barriers prevent cell state transitions, while in overly permissive chromatin (*right*) these barriers are lowered and allow easy transition to other cell states. The three scenarios are illustrated as energy state diagrams (*top*), via an example of the underlying molecular mechanisms (*center*) or as cell state transitions (*bottom*). Blue nuclei represent normal cells, while red nuclei indicate cancer cells

In a restricted state, cells may be blocked in differentiation and continue to grow with a high proliferation rate. In contrast, in permissive cells a KDM, such as KDM6A, inhibits the action of EZH2 and removes H3K27me3 marks. KDMs are often up-regulated under stress conditions, since they are responsive to signals from the tumor microenvironment. In net effect, this leads to the dominance of H3K4me3 marks and to the activation of gene expression, such as of oncogenes, even in the absence of specific stimuli. In the cell state transition diagram (Fig. 10.6, *right*, see also Fig. 8.10) the barrier between the cell states is either of medium height in normal cells, very high in restricted cells or rather low in permissive cells.

Permissive chromatin has a high rate of plasticity. This allows cancer cells to acquire easily a number of different transcriptional states, for example, shifting to alternative developmental programs, some of which can be pro-oncogenic. When such an adaptive chromatin state propagates through mitosis, a new cell clone is created that overgrows other cells due to increased fitness (Fig. 10.7). This plasticity model can be considered as the epigenetic counterpart to the genetic model of genome instability being induced by carcinogen exposure or DNA repair defects. Interestingly, in both models there are "driver" events, such as the activation of an oncogene, and "passenger" events that do not alter the fitness of the cells.

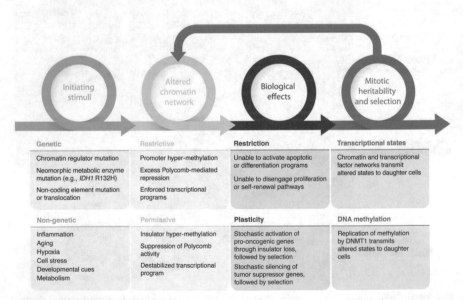

Fig. 10.7 Disruption of chromatin homeostasis in cancer. Chromatin homeostasis can be disrupted by genetic factors, such as mutations in chromatin proteins or translocation of regulatory elements, or by non-genetic factors, such as inflammation, stress or hypoxia. This can result in very permissive or restrictive chromatin networks. In permissive states oncogenic epigenetic changes may occur, such as silencing of tumor suppressor genes. Mitotically heritable, adaptive epigenetic changes will be selected and contribute to the hallmarks of cancer (Fig. 10.9)

Chromatin homeostasis is closely linked to metabolic conditions (Fig. 10.7, Sect. 13.1). Many chromatin proteins, such as DNA- and histone-modifying enzymes, use metabolites as donors and co-factors, like α-KG, the methyl donor SAM from the folate pathway and acetyl-CoA. Therefore, they are sensitive to shifts in the concentration of these metabolites. For example, mutations of IDH enzymes lead to the accumulation of 2-hydroxyglutarate inhibiting the demethylation of DNA via TET enzymes (Sect. 13.2). The resulting DNA hyper-methylation disrupts the binding of the methylation-sensitive transcription factor CTCF to insulator regions. Accordingly, the partition of the genome into discrete functional domains, in which enhancers regulate their appropriate gene targets, is disturbed. For example, reduced CTCF binding in IDH mutant gliomas causes a transcriptome profile that indicates insulator dys-function. Thus, the oncogenicity of IDH mutants seems to be primarily based on a loss of gene insulation, i.e. on compromised CTCF-mediated genome topology being an effect of epigenetic dys-regulation.

10.4 Epigenetic Mechanisms of the Hallmarks of Cancer

Some 15 years ago the concept of the "hallmarks of cancer" was introduced initially for the processes (1) sustained proliferative signaling, (2) evading growth suppressors, (3) resisting cell death, (4) enabling replicative immortality, (5) inducing angiogenesis and (6) activating invasion and metastasis, all of which are found during tumorigenesis of basically all types of cancer (Fig. 10.8). Later on, the concept was extended to 10–12 hallmarks, including "genomic instability and mutation" and "epigenomic disruption." This acknowledges the results of large-scale cancer genomics projects on both genetic and epigenetic drivers of different types of cancer. For example, they indicated high frequency of mutations in epigenetic regulators, some of which form hot spots.

The cancer genome and epigenome influence each other in a multitude of ways and can work mutually, such as in CIMP (Fig. 10.8). Both genomics and epigenomics offer complementary mechanisms to achieve similar results, such as the inactivation of tumor suppressor genes by either deletion or epigenetic silencing. For example, a *PDGFRA* oncogene gain-of-function activation that is important for achieving the hallmark "sustained proliferative signaling" may be based either on a genome mutation within the coding region of the gene or by an epimutation that disrupts insulators at the borders of the TAD carrying the gene. Similarly, the hallmark "evasion of growth suppressors" can be based on either a loss-of-function mutation of the tumor suppressor gene *CDKN2A* or by hyper-methylation of its promoter. The relative contribution of genetic and epigenetic mechanisms to the hallmarks of cancer differs between cancer types (Fig. 10.9). Interestingly, the example of the adult brain tumor glioblastoma in comparison to the childhood brain tumor ependymoma suggests that long-term tumorigenesis in adults may rather be based on genetic events, while short-term tumorigenesis has majorly an epigenetic origin.

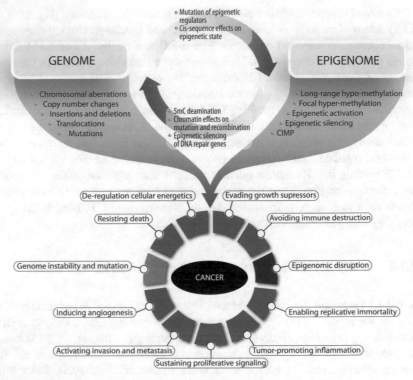

Fig. 10.8 Interplay between genome and epigenome in cancer. Changes in the genome can influence the epigenome and vice versa. This forms a network that produces genetically or epigenetically encoded variations in the phenotype that are subject to Darwinian selection for growth advantage and thus eventually achieving the hallmarks of cancer

10.5   Epigenetic Cancer Therapy

The increasing appreciation of the contribution of altered epigenetic states to the phenotype of cancer cells suggests the epigenetic therapies may have a clinical impact. This requires profound understanding how epigenetic lesions drive cancers, i.e. there is a need for conceptual and mechanistic models of cancer epigenetics (Figs. 10.4 and 10.6) in context with genetic models (Fig. 10.7). Moreover, the application of new methods, such as epigenome-wide single-cell assays (Box 2.1), combined with the selection of most appropriate human cohorts (Box 9.1) will be essential. The main arguments for epigenetic cancer therapy are that (1) genes encoding for epigenetic modifiers, such as KMTs and KDMs (Fig. 10.10), are frequent drivers in a larger range of cancer types, (2) epigenetic modifications, in contrast to genetic mutations, are largely reversible. So far, basically all molecules designed for epigenetic therapy are enzyme inhibitors. Since histone methylation marks have a far more selective function than histone

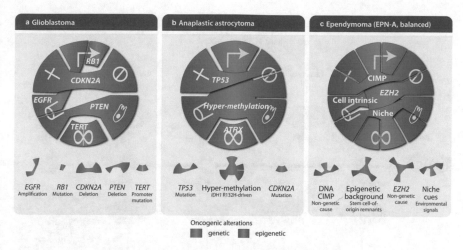

Fig. 10.9 Genetic and epigenetic mechanisms underlying the hallmarks of cancer. Both genetic (*green*) and epigenetic (*blue*) mechanisms are important factors in tumorigenesis, but their relative contribution to the hallmarks of cancer depends on the type of cancer. In glioblastoma (*a*), an adult brain tumor, most hallmarks relate to genetic drivers, while in ependymoma (*c*), a childhood tumor, primarily epigenetic effects dominate. Anaplastic astrocytoma (*b*) represents an example where both genetic and epigenetic factors contribute to the hallmarks

acetylation marks, KMT and KDM inhibitors promise to be more specific and may be less toxic than KDAC inhibitors or even DMNT inhibitors.

10.5.1 KMT Inhibitors

Several KMT inhibitors have been developed (Fig. 10.10), and those of the H3K27-KMT EZH2 (KMT6A) and the H3K79-KMT DOT1L (KMT4) are already in clinical trials. Since EZH2 is the catalytic core of the PRC2 complex that also recruits DNMTs, EZH2 inhibitors may link both epigenetic repression mechanisms. The inhibition of EZH2 results in reduced levels of H3K27me3 marks, up-regulation of silenced genes and inhibition of the growth of cancer cells with EZH2 gain-of-function mutations or overexpression.

10.5.2 KDM Inhibitors

KDMs use FAD, α-KG or Fe(II) as co-factors (Sect. 6.3) and offer in this way a number of options for their inhibition. However, the catalytic Jumonji domain of most KDMs is structurally highly conserved, which is a challenge for the design of specific KDM inhibitors (Fig. 10.10). Therefore, so far only inhibitors of the FAD-dependent H3K9-KDM LSD1 (KDM1A), GSK2879552, tranylcypromine, INCB059872 and ORY-1001, are in clinical trials.

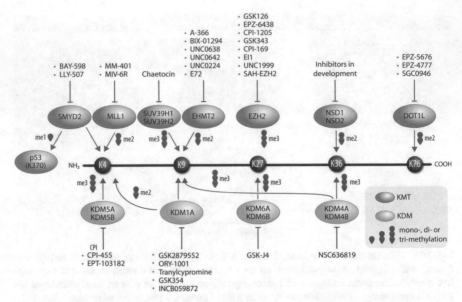

Fig. 10.10 Targeting KMT and KDM mutations. Inhibitors of KMTs (*red*) and KDMs (*blue*) are indicated that affect histone 3 lysines K4, K9, K27, K36 and K79 (*blue*). KMT inhibitors bind either within the SAM pocket, within the substrate pocket or at allosteric sites of the KMT protein. Respective inhibitors of DOT1L (KMT4) and EZH2 (KMT6A) are already in clinical trials. Chaetocin is a non-selective inhibitor of KMTs, such as SUV39H1 (KMT1A) and SUV39H2 (KMT1B). KDM1A is a FAD-dependent KDM, which can be inhibited by molecules blocking its co-factor binding site. The KDM1A inhibitors tranylcypromine, GSK2879552, INCB059872 and ORY-1001 are in clinical trials. Most KDMs carry a catalytic domain, which can be inhibited by iron-chelating molecules

10.5.3   KDAC Inhibitors

KDAC inhibitors re-activate the transcription of tumor suppressor genes, such as *CDKN1A*, by increasing histone acetylation, but they also deacetylate non-histone proteins, such as p53, and stabilize their activity. In this way, they have a wide impact on cancer cells and can induce apoptosis, cell cycle arrest and many other anti-cancer actions. Three KDAC inhibitors, vorinostat (SAHA), belinostat and romidepsin have FDA approval for treatment of different types of leukemia and more than 10 others are in clinical trials for both blood and solid tumors. However, the selectivity and detailed mechanism of action of KDAC inhibitors is still not fully understood. KDAC inhibitors are also considered for the therapy of neuronal diseases (Sect. 11.3).

10.5.4   DNMT Inhibitors

DNMT inhibitors are either nucleoside analogs that after incorporation into the DNA covalently trap DNMTs or non-nucleoside analogs that directly bind to the catalytic region of DNMTs. The molecules prevent DNA methylation leading to reduced promoter hyper-methylation and re-expression of silenced tumor suppressor genes. The two nucleoside analogs azacitidine (5-azacytidine) and decitabine (5-aza-2'-deoxycytidine) have already an approval by the FDA. Many other DNMT inhibitors are in development. Non-nucleoside analogs are less toxic, since they do not get incorporated into DNA, but they lack potency and specificity.

For the treatment of cancer mono-therapies are rarely effective, so that most cancers are treated more efficiently through the combinations of drugs. In future epigenetic therapies, in particular for solid tumors, either an epigenetic drug may be combined with another cancer treatment approach or two or more epigenetic drugs may be used in parallel. For example, the simultaneous inhibition of DNA methylation and histone deacetylation through the combined application of a DNMT inhibitor and a KDAC inhibitor should increase the chance of re-expression of, for example, tumor suppressor genes.

Key Concepts

- More than 50% of human cancers carry mutations in key chromatin proteins.
- Changes in the epigenome, such as in the patterns of DNA methylation, histone modifications and 3D chromatin structure, respectively, play a profound role in tumorigenesis.
- There are four major types of epigenetic mutations affecting cancer: DNA hyper-methylation at promoters, genome-wide DNA hypo-methylation, abnormal modification of histones and/or their recognition, and abnormal chromatin structures caused by mal-functional chromatin remodelers.
- Global DNA hypo-methylation during tumorigenesis generates chromosomal instability, reactivates transposons and causes loss of imprinting.
- Many tumors re-activate programs of fetal development, i.e. they undergo epigenetic reprograming.
- Specific genetic, environmental and metabolic stimuli disrupt the homeostatic balance of chromatin of cancer cells, which then either becomes very restrictive or permissive.
- Restrictive chromatin states prevent the induction of tumor suppressor programs or block differentiation, while permissive states allow oncogene activation or non-physiologic cell fate transitions, i.e. it has a high rate of plasticity.
- The effects of diverse oncogenic stimuli are mediated via epigenetic modulators and contribute to diverse aspects of cancer biology, such as all hallmarks of cancer.
- The relative contribution of genetic and epigenetic mechanisms to the hallmarks of cancer differs between cancer types.
- The main arguments for epigenetic cancer therapy are, that genes encoding for epigenetic modifiers are frequent drivers in a larger range of cancer types, and that epigenetic modifications are largely reversible.

(continued)

Key Concepts (continued)
- Basically all molecules designed for epigenetic therapy are inhibitors of KMTs, KDMs, KDACs and DMNTs.
- For the treatment of cancer, mono-therapies are rarely effective, so that most cancers are treated more effectively through the combinations of drugs.

Additional Reading

Feinberg AP, Koldobskiy MA, Gondor A (2016) Epigenetic modulators, modifiers and mediators in cancer aetiology and progression. Nat Rev Genet 17:284–299

Flavahan WA, Gaskell E, Bernstein BE (2017) Epigenetic plasticity and the hallmarks of cancer. Science 357:eaal2380.

Hitchins MP (2015) Constitutional epimutation as a mechanism for cancer causality and heritability? Nat Rev Cancer 15:625–634

Pfister SX, Ashworth A (2017) Marked for death: targeting epigenetic changes in cancer. Nat Rev Drug Discov 16:241–263

Shen H, Laird PW (2013) Interplay between the cancer genome and epigenome. Cell 153:38–55

Suva ML, Riggi N, Bernstein BE (2013) Epigenetic reprograming in cancer. Science 339:1567–1570

Timp W, Feinberg AP (2013) Cancer as a dysregulated epigenome allowing cellular growth advantage at the expense of the host. Nat Rev Cancer 13:497–510

Chapter 11
Neuroepigenetics

Abstract Genome-wide DNA methylation patterns significantly change during brain development and maturation and are the basis for neuronal plasticity. Widespread methylome reconfiguration, such as non-CG methylation (mCH), occurs in neurons, but not in glial cells, during fetal to young adult development and becomes the dominant form of methylation in the human neuronal genome. In parallel, during brain development, there is an increase of 5hmC marks and possibly CpG demethylation in particular at gene bodies.

Rapid and dynamic methylation and demethylation of specific genes in the brain may play a fundamental role in learning, memory formation, and behavioral plasticity. MeCP2 is the best-characterized methyl-binding transcription factor and is involved both in gene activation and repression. MeCP2 is highly expressed in the brain and an important component of neuronal chromatin, where it reduces – via replacing the linker histone H1 – the chromatin repeat length to 165 bp. Mutations in the *MECP2* gene are the mechanistic basis of the autism spectrum disorder Rett syndrome. In addition, also the histone acetylation level in neurons contributes to the cell's proper function. Accordingly, KDAC inhibitors offer an effective therapy in neurodegenerative diseases, such as Rubinstein-Taybi syndrome, Friedreich ataxia and Huntington disease, in which the homeostasis of this epigenetic mark is disturbed. The transcription factor REST (RE1-silencing transcription factor) acts as DNA-binding platform for a large number of chromatin modifiers, such as KDACs, KMTs and KDMs, and primarily mediates silencing of its neuronal target genes. Dys-regulation of REST provides insight into epigenetic processes in the context of Alzheimer and Huntington diseases.

In this chapter, we will describe the field of neuroepigenetics and will provide mechanistic explanations for the contribution of epigenetics to neurodevelopmental and neurodegenerative diseases.

Keywords Neurodevelopment · DNA methylation · MeCP2 · Rett syndrome · histone acetylation · KDAC inhibitors · REST · neurodegenerative diseases

11.1   Epigenetics of Neuronal Development and Memory

The frontal cortex region of the brain plays a key role in behavior and cognition and requires a coordinated interaction of neuronal and non-neuronal cells, such as supporting glial cells. The highly controlled process of development and maturation creates the physical structure of the brain. It starts during embryogenesis and continues until the third decade of life. In parallel, after birth there is first a burst of synaptogenesis and then a pruning of unused synapses during adolescence. This is the basis for experience-dependent plasticity and learning in children and young adults. Its disruption can lead to behavioral alterations and neuropsychiatric disorders.

In other developmental programs of the human body (Sect. 8.4), also neurodevelopment is controlled by precise epigenetic patterns, such as DNA methylation and histone modifications (Sect. 11.3). Thus, neuroepigenetics uses the same mechanisms as all other human tissues and cell types. However, neurons do not divide, i.e. they do not propagate epigenetic information to successive cell generations. In contrast, in neurons epigenetic mechanisms are used for information storage and the regulation of circuits. The failure of these processes can disturb or even disrupt basic network-related cognitive function and thus may contribute to neurodegenerative diseases.

DNA methylation plays not only a central role during fetal development of the brain but also during its maturation after birth. Also in the brain DNA methylation typically silences gene expression, but in addition is an essential mediator of memory acquisition and storage. For example, in mice the knockout of *Dnmt1* and/or *Dnmt3a* results in the loss of long-term potentiation (LTP) and consecutively in deficits in learning and memory, i.e. without active DNA methylation these information storage processes do not work.

Most epigenome studies focus on 5mC marks at CpGs (mCG), but – in addition to ES cells – adult neurons have the special property to carry significant levels of mCH marks (H = A, C or T, Sect. 4.1), some 70% of which are mCA marks. While mCH marks are hardly detectable in fetal brain, they increase during early post-natal development to a maximum of 1.5% of all CH dinucleotides at the end of adolescence (Fig. 11.1, *left*). mCH levels rise most rapidly during the primary phase of synaptogenesis, i.e. within the first 2 years after birth, and correlate with the increase in synapse density.

Since the frontal cortex develops post-natally in response to various inputs from the environment, neuronal-specific DNA methylation leads to changes in gene expression and synaptic development that are dependent on sensory input. During development, DNMT3A sets these mCH marks, in particular at CA dinucleotides. In the adult brain mCH marks are by number even more abundant than mCG marks (the average methylation percentage at CH is clearly lower than at CG, but CpGs are more rare). This highlights mCA as a major epigenetic mark having repressive function on gene regulation in the maturing brain.

The transcription factor MeCP2 (Sect. 4.1) recognizes with high affinity mCA marks. Therefore, genes that acquire mCA enrichment during neurodevelopment recruit MeCP2 and are repressed in their transcriptional initiation and elongation (Fig. 11.2). In general, mCA marks may serve as a docking platform for

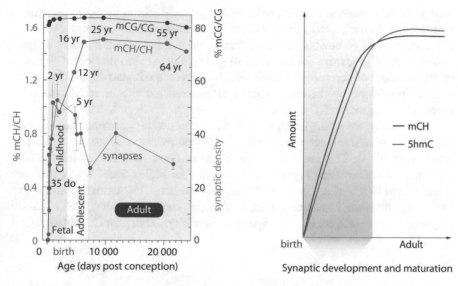

Fig. 11.1 Changes of the neuronal methylome during brain development and maturation. Measured on a single-base resolution (Sect. 2.1), the levels of mCH (*left*) and 5hmC (*right*) accumulate in neurons of the human frontal cortex after birth, which coincides with active synapse development (indicated as synaptic density per 100 mm³, *left*) and maturation. mCH marks cause gene silencing, while 5hmC marks are found at active genes. (Data are based on Lister et al. 2013.)

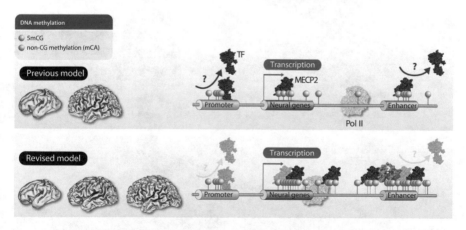

Fig. 11.2 Neuronal gene regulation via methylation. Previously, it was assumed that mCG marks block transcription factor binding at discrete regulatory elements in the genome, such as enhancers, or attract MeCP2 (*top*). In both cases transcription is repressed. This model does not reflect the epigenomic differences in the brains of newborns and adults. A revised model (*bottom*) highlights the accumulation of mCH marks, which is initiated at birth and continues throughout adolescence. In the adult brain the high-affinity interaction between mCA and MeCP2 (as well as possibly other mCH binding proteins) has specific effects on transcriptional initiation and elongation of genes that are enriched in mCA

methyl-binding proteins, for example, at the genomic region of the brain-derived neurotrophic factor (*BDNF*) gene. This mediates distinct functions in the brains of adult versus that of newborns. Accordingly, the number of mCA marks at gene bodies is more predictive for the level of gene silencing than that of mCG marks or any measure of chromatin accessibility. Moreover, the impact of mCH marks is further highlighted by the observation that between individuals mCH marks are more conserved than mCG marks.

Different regions of the brain, such as frontal cortex, hippocampus and cerebellum, show significant age-dependent increase of oxidized forms of 5mC, like 5hmC (Fig. 11.3). This increase is specific to neurons, so that they have – with exception of the fertilized zygote – in the adult a far higher 5hmC levels than any other human tissue of cell type. TET enzymes mediate 5mC oxidation and active DNA demethylation in neurons. Based on mouse knockout studies, TET1 is most important for neurogenesis and synaptic plasticity. Thus, 5mC oxidation and

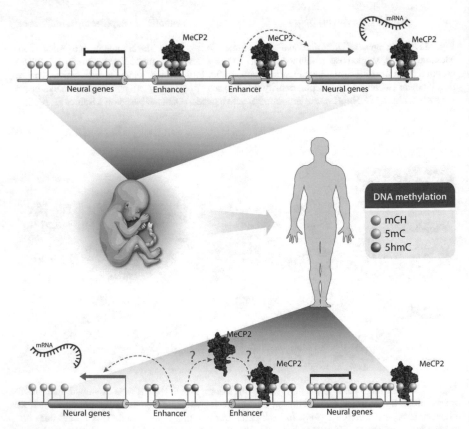

Fig. 11.3 The neuronal methylome during brain development. 5hmC marks label enhancers in the fetal brain (*top*) that will become demethylated and active in the adult brain (*bottom*). The DNA-binding pattern of methyl-DNA-binding proteins, which have different affinity for 5hmC and 5mC, can change affecting neuronal transcription programs

possibly DNA demethylation are central epigenetic mechanisms of brain development and maturation.

In the adult brain the epigenetic mark 5hmC can recruit MeCP2 and other methyl-binding proteins. Their abundance of these transcription factors mostly positively correlates with gene expression. This fits with the observation that the *MECP2* mutation R133C, which occurs in some forms of the Rett syndrome (Sect. 11.2), specifically disrupts the ability of the protein to bind to 5hmC resulting in target gene silencing. Moreover, the transcriptional repressor THA11 (THAP domain-containing 11), which has a central role in embryogenesis, is a brain-specific 5hmC reader.

In general, enrichment of 5hmC and depletion of 5mC both increase transcriptional activity and chromatin accessibility (Fig. 11.3). This also implies that 5mC oxidation may be the key epigenetic mechanism in controlling gene expression in the context of synaptic plasticity, which is important for learning and memory. The latter involves effects on LTP, excitability and activity-dependent synaptic scaling of neurons. Thus, via controlling ion channels, receptors and trafficking mechanisms the epigenome has the capacity to both sense extra-cellular signals reaching neurons and to control their output.

Taken together, neurons have epigenetic marks, such as most mCGs and mCHs, that – after they have been set during development – stabilize the neuronal phenotype over the lifespan. For this reason, neurodevelopmental disorders are generally thought to be irreversible (Sects. 11.2 and 11.3). However, a plastic nature of the dynamic part of the neural epigenome via 5hmC and demethylation has major implications for a possible epigenetic reprograming in the context of these disorders.

11.2 MeCP2 and the Rett Syndrome

MeCP2 belongs to the family of intrinsically disordered proteins, which are characterized by a low level of secondary structure allowing them to interact with different types of macromolecules, such as other proteins, DNA and RNA. Central to the primary structure the MeCP2 protein is its DNA-binding domain that includes a MBD, but also contains several other structures, such as AT hook motifs (Fig. 11.4). Accordingly, a specific target of MeCP2 is methylated DNA, in particular CA dinucleotides, but the protein also interacts with a wide range of proteins, such as heterochromatin protein HP1, the co-repressors NCOR1 and SIN3A and the KAT CREBBP.

MeCP2 gets rapidly ubiquitinated, which limits its half-life to only 4 hours. The protein is ubiquitously expressed in basically all human tissues and cell types, but it shows highest expression in the brain, in particular in neurons. Accordingly, in neuronal chromatin MeCP2 is a very abundant protein appearing in average in every two nucleosomes (Fig. 11.5). MeCP2 and the linker histone H1 compete for binding to the nucleosome. When MeCP2 levels rise during the neuronal

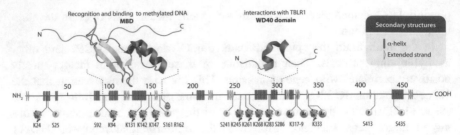

Fig. 11.4 Structure of MeCP2. The tertiary structure of MeCP2's MBD and WD40 domain as well as the predicted secondary structure along with sites of post-translational modification of the whole protein are shown

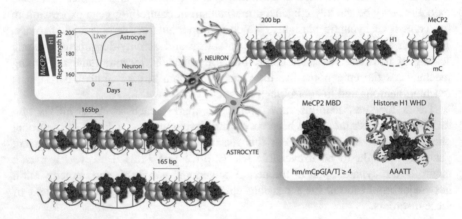

Fig. 11.5 MeCP2 binding to chromatin of neurons and astrocytes. Astrocytes have lower MeCP2 levels than neurons resulting in a regular chromatin repeat length of 200 bp compared to 165 bp for neurons. In neurons MeCP2 is evenly distributed throughout the chromatin, binds to methylated DNA sites and replaces the linker histone H1, which decreases the repeat length. The change of chromatin repeat length during development (in days, observed in a mouse model) before and after birth are indicated for astrocytes, neurons and as a reference for liver. A higher proportion of MeCP2 in relation to histone H1 results in a shorter repeat length (insert *top left*)

development, the protein replaces histone H1, which causes the reduction of the chromatin repeat length (i.e. the distance from the center of a nucleosome to the center of its neighbor) from 200 bp to 165 bp (Fig. 11.5). In this configuration, MeCP2 is part of tightly folded heterochromatin structures suggesting that the protein mainly functions as a transcriptional repressor.

However, there are a growing number of MeCP2 interaction partners that stimulate gene expression. This indicates that MeCP2 rather is a transcriptional regulator that – depending on its interaction partners – binds to genomic regions of activated or repressed genes. Furthermore, MeCP2 has opposite roles when binding to promoter regions or gene bodies (Fig. 11.6). DNA methylation at TSS regions recruits MeCP2 in complex with co-repressor proteins and KDACs and results in transcriptional repression. In contrast, the DNA of gene bodies of

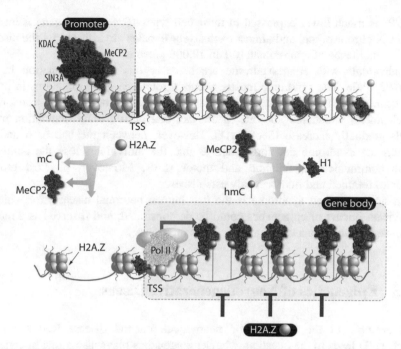

Fig. 11.6 MeCP2-mediated neuronal gene expression. MeCP2 binds at methylated promoter regions together with co-repressor Sin3A and KDACs and represses gene transcription (*top left*). DNA demethylation at these genomic regions increases chromatin accessibility through nucleosome loss. This is caused by insertion of the histone variant H2A.Z and also results in loss of MeCP2 (*bottom left*). Finally, this allows the assembly of the basal transcriptional machinery on the TSS and initiates gene transcription. Interestingly, at gene bodies (*bottom right*) of transcribed genes DNA is methylated and recruits MeCP2 preventing H2A.Z from binding

transcribed genes is methylated and recruits MeCP2 binding, which in turn prevents the binding of the histone variant H2A.Z (Sect. 3.3). Thus, the genomic location of MeCP2 binding is of critical importance of its function.

Chromatin modifiers play central roles in cognitive disorders. At least seven proteins are known to be mutated in X chromosome-linked intellectual disabilities. These proteins, such as MeCP2, are either methyl-binding proteins or methyl-modifying enzymes. The disruption of the *MECP2* gene leads to a special form of autism, referred to as Rett syndrome. It shows some overlap with autism, but not all children with Rett syndrome display the complete set of autism features. For example, in contrast to autism, at an age of five girls with Rett syndrome are socially very interactive and rarely have problems with the movement coordination of small muscles.

More than 95% of all cases of the Rett syndrome can be explained by mutations within the *MECP2* gene, i.e. it is largely a monogenetic disease. This means that the mechanistic understanding of the Rett syndrome is based on dysfunctional MeCP2 proteins in neurons. In addition, also functional alternations of MeCP2 in astrocytes and in microglia affect the disease phenotype, although

MeCP2 is much lower expressed in these cell types. The *MECP2* gene is located on the X chromosome, and almost exclusively females are affected by the disease with an incidence of approximately 1 in 10,000 persons.

Individuals with Rett syndrome are heterozygous for the mutation in the *MECP2* allele. The syndrome mostly occurs through de novo mutations in paternal germline cells. Affected females have an apparent normal early post-natal development, which may be due to the fact that during brain maturation mCH levels gradually increase (Sect. 11.1). However, between the age of 6 and 18 months the syndrome develops, such as that the individuals lose the ability to retain communication function and motor skills. Moreover, physical growth becomes retarded and microcephaly establishes.

In summary, Rett syndrome is the first human neuronal disease, for which a significant impact of epigenetics could be demonstrated, and it serves as a master example for the impact of neuroepigenetics on disease.

11.3   Epigenetics of Neurodegenerative Diseases

The example of the monogenetic neurodevelopmental disease Rett syndrome (Sect. 11.2) leads to the question, whether epigenetics plays also a role in complex (i.e. multi-factorial) disorders of the nervous system, such as neurodegenerative diseases.

MeCP2 is involved in the secretion of the neurotransmitters GABA, dopamine and serotonin from respective neurons and controls the number of synapses of glutamatergic neurons. This is, at least in part, mediated by the interaction of MeCP2 with BDNF (Sect. 11.1), which acts as a modulator on glutamatergic and GABAergic synapses. Thus, the tightly regulated expression of MeCP2 is critical for neuronal homeostasis. Since both too high as well as too low MeCP2 protein levels trigger opposite effects in synaptic transmission, dys-regulation of MeCP2 protein expression contributes to many neuro-pathological disorders, such as Alzheimer disease, Huntington diseases, schizophrenia and epilepsy (Table 11.1). In parallel, aberrant DNA methylation has been observed in some autism spectrum disorders, Alzheimer disease, epilepsy and schizophrenia. In general, epigenetic mechanisms may be particularly relevant to complex diseases with low genetic penetrance that use epigenetic mechanisms of development and learned behavior (Sect. 11.1), such as drug addiction, post-traumatic stress disorder (PTSD), epilepsy and schizophrenia.

Histone acetylation is the best-understood epigenetic modification and most tightly associated with memory formation. Accordingly, several neurodegenerative diseases involve disruptions in the KAT/KDAC balance, i.e. patients with these disorders have abnormal histone acetylation levels (Fig. 11.7). A master example is the Rubinstein-Taybi syndrome, which is characterized by short stature, mental retardation, moderate to severe learning difficulties, distinctive facial features, and broad thumbs and first toes. Rubinstein-Taybi syndrome is a monogenetic disease

Table 11.1 Epigenetic mechanisms of neuropathological disorders. Neuronal disorders and functions that are affected by epigenetics are listed, such as mutations or overexpression of MeCP2, aberrant DNA methylation and/or histone modifications

Function or disorder	Mechanism(s) implicated
Rett syndrome	*MECP2* mutations
Autism spectrum disorders	MeCP2 overexpression/increased dosage, aberrant DNA methylation
Alzheimer disease	MeCP2 decrease, histone modifications, aberrant DNA methylation
Parkinson disease	Loss of MeCP2
Huntington disease	MeCP2 dys-regulation
Fragile X syndrome	aberrant DNA methylation
Rubinstein-Taybi syndrome	KAT deficiency
Friedreich ataxia	Reduced histone acetylation
Angelman syndrome	Genomic imprinting (DNA methylation)
Addiction and reward behavior	MeCP2 decrease, histone modifications, DNA methylation, miRNAs
PTSD	Histone modifications, DNA methylation
Depression and/or suicide	DNA methylation
Schizophrenia	Increased MeCP2 binding, histone methylation, aberrant DNA methylation
Epilepsy	MeCP2 up-regulation, histone modifications, aberrant DNA methylation

that is based on mutations of the *CREBBP* gene, which encodes for a KAT (Sect. 6.1). The resulting low histone acetylation levels may be counterbalanced by the inhibition of KDACs, i.e. KDAC inhibitors (Table 11.2) are a therapeutic option for patients with the syndrome.

Another example of a monogenetic neurodegenerative disease is Friedreich ataxia that results from the degeneration of nervous tissue in the spinal cord, in particular in sensory neurons that are essential for directing muscle movement of arms and legs. In this disorder the expansion of a triplet repeat region within an intron of the frataxin (*FXN*) gene leads to the loss of H3ac and H4ac marks and gain of H3K9me3 marks, heterochromatin formation and finally transcriptional silencing of the gene (Fig. 11.7). In a mouse model of this disease, KDAC inhibitors increased H3 and H4 acetylation and corrected the *FXN* expression deficiency. This suggests that KDAC inhibitors may be suited for the treatment of Friedreich ataxia.

Similarly, the mental retardation associated with the fragile X syndrome is caused by a CGG-triplet repeat expansion in the 5′-UTR of the human fragile X mental retardation 1 (*FMR1*) gene, which leads to extensive DNA methylation at CpGs close to the gene's TSS and gene silencing (Fig. 11.7). Also in this case a treatment with KDAC inhibitors resulted in re-activation of *FMR1* expression.

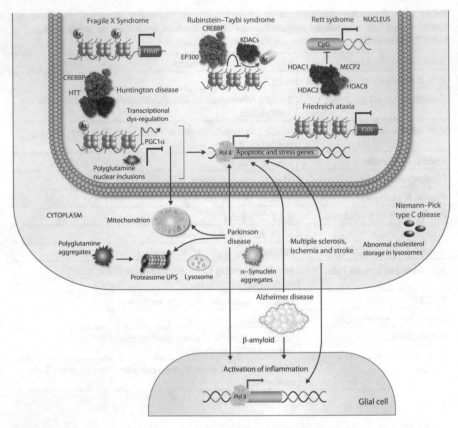

Fig. 11.7 Role of histone acetylation in neurodegenerative diseases. The level of histone acetylation depends on the balance of KATs and KDACs and is related to several neurodegenerative diseases. For example, decreased acetylation activity of the KAT CREBBP is associated with the Rubinstein-Taybi syndrome and polyglutamine diseases, such as Huntington disease. The transcriptional silencing of the *FXN* gene in Friedreich ataxia or of the *FMR1* gene in fragile X syndrome can be released by KDAC inhibition. Other examples of neuronal disorders are schematically depicted and their possible treatment by KDAC inhibitors is indicated

In particular SIRT1 inhibitors were able to increase acetylation and decrease methylation of histones at the *FMR1* locus. Finally, Huntington disease is based on polyglutamine repeats in the 5′-coding region of the huntingtin (*HTT*) gene causing progressive motor and cognitive decline. The disease involves perturbations in many aspects of neuronal homeostasis, such as abnormal histone acetylation and chromatin remodeling as well as aberrant interactions of the HTT protein (Fig. 11.7). In different animal models of Huntington disease KDAC inhibitors showed a neuroprotective effect.

Targeting histone acetylation via the application of KDAC inhibitors may also provide benefit for the treatment of complex neuronal diseases, such as Alzheimer disease and Parkinson disease as well as depression, schizophrenia, drug addiction

Table 11.2 KDAC inhibitors for the treatment of neuronal disorders. Some KDAC inhibitors are since longer time approved by the FDA for blood cancers treatment (Sect. 10.5). In addition, KDAC inhibitors are also used in psychiatry and neurology as mood stabilizers and anti-epileptics. Furthermore, KDAC inhibitors are studied as for a treatment of neurodegenerative diseases. The compounds may have therapeutic potential in acute brain injury, stroke, Alzheimer disease, Huntington disease and Parkinson disease

KDAC inhibitor	Drug target	Stage of development	Disease models tested
Valproic acid	Class I & II KDACs	FDA approved in 1986	Alzheimer disease, Huntington disease, stroke and global ischemia, approved as mood stabilizer and anti-epileptic
Phenylbutyrate	Class I & II KDACs	FDA approved in1996	Alzheimer disease and Huntington disease
Suberoylanilide hydroxamic acid (vorinostat)	Class I & II KDACs	FDA approved in 2006	Alzheimer disease, Huntington disease and stroke
Nicotinamide (SIRT inhibitor)	Class III KDACs	Phase III	Alzheimer disease and Huntington disease
Entinostat (MS275)	Class I KDACs	Phase II	Alzheimer disease and stroke
Trichostatin A	Class I & II KDACs	Preclinical	Huntington disease, stroke and global ischemia
AGK2 (SIRT2 inhibitor)	Class III KDACs	Preclinical	Huntington disease and Parkinson disease
Sodium butyrate	Class I & II KDACs	Preclinical	Alzheimer disease, Huntington disease, stroke, brain atrophy and depression

and anxiety disorders (Table 11.2). For example, in mouse models the KDAC inhibitor sodium butyrate showed anti-depressant effects. Moreover, KDAC inhibitor treated animals showed induced sprouting of dendrites, an increased number of synapses, and re-established learning behavior and access to long-term memories. This suggests a possible wide application for KDAC inhibitors in the therapy of cognitive disorders. Although KDAC inhibitors got initially FDA approval for the treatment on different types of cancer (Sect. 10.5), valproic acid (Table 11.2) was also approved as mood stabilizer and anti-epileptic.

KDAC inhibitors also widely affect gene expression in the immune system (Sect. 10.5) and showed to be effective in the treatment of inflammation and neuronal apoptosis (Fig. 11.7). Animal models of ischemia-induced brain infarction indicated that KDAC inhibitors have anti-inflammatory and neuroprotective effects and can be used for the treatment of stroke (Table 11.2).

Mechanistically, several pathways can explain how histone acetylation is triggered by neuronal activity (Fig. 11.8, *left*). For example, potassium channel-mediated depolarization of neurons results in increased acetylation of histone H2B, while induction of dopaminergic, cholinergic and glutamatergic pathways

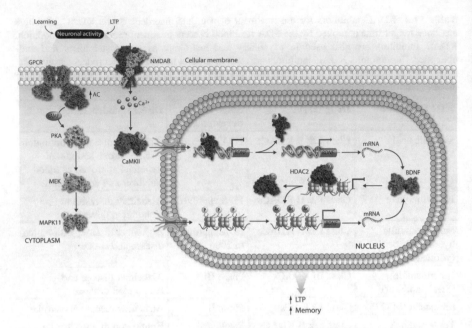

Fig. 11.8 Histone acetylation due to neuronal activity. Neuronal activity signals to the nucleus induce epigenetic modifications that sustain LTP and memory (*left*). LTP and learning stimulate G protein-coupled receptors (GPCRs), which in turn activate adenylyl cyclase (AC) that converting ATP to cAMP (*center*). In the following, cAMP stimulates protein kinase A (PKA), which phosphorylates MEK leading to the phosphorylation of MAPKs. Members of the MAPK family can directly phosphorylate histones, which subsequently triggers histone acetylation. Alternatively, neuronal activity can also lead to cellular depolarization via calcium influx, which activates calmodulin and results in the phosphorylation-dependent activation of CaMKII (*right*). CaMKII phosphorylates MeCP2, which induces the dissociation of the transcription factor from the *BDNF* promoter region allowing the transcription of the latter. BDNF activates NO synthase leading to nitrosylation of HDAC2. Nitrosylated HDAC2 dissociates from chromatin resulting in increased histone acetylation at the promoter region of its target genes (including *BDNF*)

causes acetylation of H3K14. These pathways involve the activation of extra-cellular regulated kinase (ERK), which is a member of the mitogen-activated protein kinase (MAPK) pathway and directly phosphorylates histones (Fig. 11.8, *center*). For example, the resulting H3S10 marks specifically induce H3K14 acety-lation. Alternatively, cellular depolarization via calcium influx can affect histone acetylation levels through the dissociation of HDAC2 from chromatin (Fig. 11.8, *right*). In this pathway, calcium influx activates calmodulin and results in the phosphorylation-dependent activation of calcium/calmodulin-dependent kinase II (CaMKII). CaMKII phosphorylates MeCP2, which dissociates from the *BDNF* promoter region and releases the silencing of the gene. BDNF activates nitric oxide (NO) synthase, which leads to the nitrosylation of HDAC2 and dissociation of the protein from chromatin. Thus, a burst of neuronal activity starts a positive-feedback loop containing HDAC2 and BDNF that induce histone acetylation-mediated, self--sustaining gene expression programs, which are the basis of synaptic plasticity and memory.

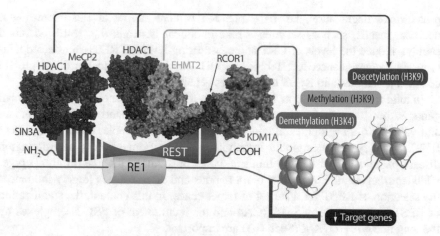

Fig. 11.9 Target gene silencing mechanism of the REST complex. Genomic binding sites of the transcription factor REST contain a 21–23 bp recognition motif, referred to as RE1. REST mediates silencing of its target genes by forming close to the TSS regions of its target genes a complex with the co-repressors RCOR1 and SIN3A, the KDACs HDAC1 and HDAC2, MeCP2, the KMT EHMT2 (KMT1C) and the KDM LSD1 (KDM1A), which is able to deacetylate and methylate H3K9 and demethylate H3K4me2. This results in local tightening of chromatin and prevents the access of the basal transcriptional machinery

Neurodegenerative diseases have different underlying causes and pathophysiology, but they all involve impaired cognition, neuronal death and the dysregulation of the transcription factor REST (also known as NRSF). REST is a gene-silencing transcription factor that has an effect on both acetylation and methylation levels of histones. REST is the DNA-binding platform of a large protein complex containing the co-repressors RCOR1 (Sect. 6.3) and SIN3A, the KDACs HDAC1 and 2, MeCP2, the KMT EHMT2 (KMT1C) and the KDM LSD1 (KDM1A) (Fig. 11.9). In addition, the REST complex can also recruit the DNA methylation machinery. The REST complex binds RE1 (restrictive element 1) sites that locate in the vicinity of TSS regions of REST target genes. This leads to deacetylation and methylation of local nucleosomes at position H3K9 and demethylation at H3K4me2 marks. Thus, the local chromatin at TSS regions of REST target genes gets very effectively silenced by a number of different mechanisms. The binding of REST to RE1 sites is stabilized by SMARCA4 (also called BRG1), a member of the SWI/SNF chromatin remodeling complex (Sect. 7.2), which recognizes via its bromodomain H4K8ac marks.

REST is widely expressed throughout embryogenesis, but at the end of neuronal differentiation it gets down-regulated, in order to acquire the neuronal phenotype. In pluripotent stem cells and neural progenitors, REST actively represses a large number of neuron-specific genes, which play a role in synaptic plasticity and structural remodeling during development, such as synaptic vesicle proteins, ion channels, neurotransmitter receptors and miRNAs regulating networks of non-neuronal genes. In iPS cells, neural progenitors and cancer cells the abundance of REST is regulated by ubiquitin-based proteosomal degradation. In differentiated neurons phosphorylation of REST by casein kinase 1 (CK1) primes the protein for ubiquitination and proteosomal degradation, i.e. its expression is kept at low levels. Interestingly, neuronal insults, such as seizures, stroke or global ischemia (i.e. restricted blood supply), reduce the levels of CK1 and cause rising levels of REST in affected neurons. As a consequence, REST binds to RE1 sites and the multi-functional protein complex assembles and REST target genes get silenced (Fig. 11.9).

In total there may be up to 2,000 primary REST target genes within the human genome, but the set of active REST targets is cell type- and context-dependent and varies with developmental and disease stage. For example, there are different REST target genes after global ischemia, in the prefrontal cortex of Huntington disease patients and of healthy humans. It can be assumed that the different epigenetic landscape in the respective brain regions and disease states largely influences the selection of REST for a subset of target genes. In this context, the stabilization of REST to RE1 sites by SMARCA4 and the recruitment of PRC2 complexes by the long ncRNA *HOTAIR* (Sect. 6.3) are important.

Since most cases of Alzheimer disease are sporadic and develop over time, there is a significant contribution of environmental factors to the onset of this neurodegenerative disorder. While in neurons of the prefrontal cortex and the hippocampus of the healthy aging brain REST silences genes involved in apoptosis and oxidative stress (Fig. 11.10, *top left*), this is lost in patients with mild or severe cognitive impairment and Alzheimer disease (Fig. 11.10, *top right*). During normal aging, REST is activated, for example, by the WNT/β-catenin pathway and mediates neuroprotection via silencing those of its target genes, which are involved in neuronal death, such as FAS, TRADD and BAX, as well as genes involved in Alzheimer disease pathology, such as PSEN1, PSEN2 and PSENEN.

With the occurrence of mis-folded proteins that are characteristic for the onset of Alzheimer disease, such as Aβ and tau, the rate of autophagy increases. Under these conditions autophagosomes engulf not only Aβ and tau complexes but also REST, which results in depletion of the transcription factor from the nucleus. The loss of REST results in an increase in expression of previously repressed genes being involved in oxidative stress and neuronal death. Thus, REST deprivation increases the loss of neurons in the respective brain regions and promotes the progression of Alzheimer disease.

In neurons and/or glial cells of the striatum (one of the nuclei in the subcortical basal ganglia of the forebrain) of healthy wild type mice HTT binds to REST and keeps it in the cytoplasm (Fig. 11.10, *bottom left*). Under these conditions REST cannot silence its target genes, such as *BDNF*, that are important for neuronal activity and survival, i.e. the genes stay expressed. In contrast, in a Huntington disease mouse model the *BDNF* gene is not expressed in these cells, because the polyglutamine expansion of HTT protein disables the binding of REST (Fig. 11.10, *bottom right*). Accordingly, REST stays in the nucleus and silences *BDNF* expression. It can be assumed that similar mechanisms apply in healthy humans and Huntington disease patients.

Taken together, neuroepigenetics is an emerging field showing potential to guide the design of novel therapeutic strategies for ameliorating the cognitive deficits and neurodegeneration.

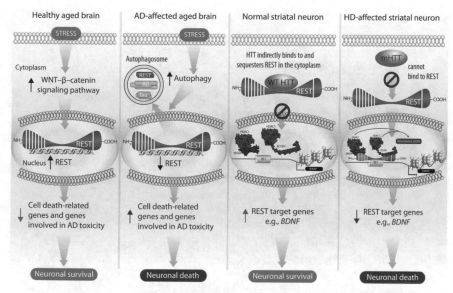

Fig. 11.10 REST in neurodegenerative diseases. In the healthy aged brain, stress increases WNT/β-catenin signaling in the hippocampus and the frontal cortex resulting in the nuclear translocation of REST and binding to RE1 sites of its target genes (*top left*). This promotes neuroprotection through the repression of proteins involved in neuronal death and in Alzheimer disease (AD) pathology. In contrast, in brains of Alzheimer patients, oxidative stress activates autophagy, in the context of which REST is engulfed (*top right*). Reduced levels of REST in the nucleus cause an increase in expression of genes that are involved in oxidative stress and neuronal death. In normal striatal neurons or glial cells of healthy individuals HTT binds and keeps REST in the cytoplasm, so that those of its target genes that are important for neuronal activity and survival stay expressed (*bottom left*). In contrast, in the respective cells of Huntington disease (HD) patients mutant HTT (mHTT) is unable to bind and sequester REST, so that it stays in the nucleus and silences its target genes, thus promoting neuronal death

Key Concepts

- Neurodevelopment is controlled by precise epigenetic patterns, such as DNA methylation and histone modifications.
- Genome-wide DNA methylation patterns significantly change during brain development and maturation and are the basis for neuronal plasticity.
- In neurons epigenetic mechanisms are used for information storage and the regulation of circuits, the failure of which can disturb or even disrupt basic network-related cognitive function and may contribute to neurodegenerative diseases.
- Widespread methylome reconfiguration, such as mCH, occurs in neurons during fetal to young adult development and becomes the dominant form of methylation in the human neuronal genome.
- mCA as a major epigenetic mark having repressive function on gene regulation in the maturing brain.

(continued)

Key Concepts (continued)

- Rapid and dynamic methylation and demethylation of specific genes in the brain may play a fundamental role in learning, memory formation, and behavioral plasticity.
- 5mC oxidation and possibly DNA demethylation are central epigenetic mechanisms of brain development and maturation.
- Mutations in the *MECP2* gene are the mechanistic basis of the autism spectrum disorder Rett syndrome.
- Rett syndrome is the first human neuronal disease, for which a significant impact of epigenetics could be demonstrated, i.e. it serves as a master example for the impact of neuroepigenetics on disease.
- The histone acetylation level in neurons contributes to the cell's proper function.
- KDAC inhibitors offer an effective therapy in neurodegenerative diseases, in which the homeostasis of this epigenetic mark is disturbed. For example, the KDAC inhibitor valproic acid is approved as mood stabilizer and anti-epileptic.
- The transcription factor REST acts as DNA-binding platform for a large number of chromatin modifiers and mediates silencing of its neuronal target genes.
- Dys-regulation of REST provides insight into epigenetic processes in the context of both Alzheimer and Huntington diseases.
- Neuroepigenetics is an emerging field showing potential to guide the design of novel therapeutic strategies for ameliorating the cognitive deficits and neurodegeneration.

Additional Reading

Ausio J, Martinez de Paz A, Esteller M (2014) MeCP2: the long trip from a chromatin protein to neurological disorders. Trends Mol Med 20:487–498

Gräff J, Tsai LH (2013) Histone acetylation: molecular mnemonics on the chromatin. Nat Rev Neurosci 14:97–111

Hwang JY, Aromolaran KA, Zukin RS (2017) The emerging field of epigenetics in neurodegeneration and neuroprotection. Nat Rev Neurosci 18:347–361

Kazantsev AG, Thompson LM (2008) Therapeutic application of histone deacetylase inhibitors for central nervous system disorders. Nat Rev Drug Discov 7:854–868

Klein HU, De Jager PL (2016) Uncovering the role of the methylome in dementia and neurodegeneration. Trends Mol Med 22:687–700

Lister R, Mukamel EA, Nery JR et al (2013) Global epigenomic reconfiguration during mammalian brain development. Science 341:1237905

Chapter 12
Epigenomics of Immune Function

Abstract Hematopoiesis, i.e. the life-long regeneration of blood cells, produces far more cells than any other tissue of the human body. The self-renewal and differentiation of HSCs is regulated by number of extrinsic and intrinsic factors, such as signal transduction pathways stimulated by growth factors, as well as transcription factors and chromatin modifiers. Most of the some 100 different blood cell types belong to the immune system and differ in their epigenetic programing, in particular at cell-specific enhancer regions, that results in sustained changes in gene expression and cell physiology.

Not only the cells of the adaptive immune system, B and T cells, but also monocytes, natural killer (NK) cells, monocytes and macrophages of the innate immune system have a memory function, referred to as "trained immunity." This memory is based on epigenetic changes, such as DNA methylation and histone modifications, and monitors the close relationship between immune challenges and the effects on chromatin. Immune-mediated diseases, such as autoimmune diseases and inflammation, are clinically heterogeneous but all develop from the interplay of genetic susceptibility and environmental or lifestyle factors, i.e. the balance between genetics and epigenetics. Accordingly, epigenetic profiling of immune cell subsets is an important tool for the understanding and possibly therapy of these diseases, for example, by small-molecule inhibitors of chromatin modifiers.

In this chapter, we will discuss the main chromatin modifiers and transcription factors controlling hematopoiesis as well as the role of cell-specific enhancers in this differentiation process. We will present epigenetic profiling as an important tool for a molecular description of heath and disease in the immune system. Moreover, we will outline the potential of small-molecule inhibitors of chromatin modifiers in the therapy of immune diseases and in particular in cancer immunotherapy.

Keywords Hematopoiesis · HSCs · epigenetic profiling poised enhancers · inflammation · autoimmunity · small-molecule inhibitors · cancer immunotherapy

© Springer Nature Singapore Pte Ltd. 2018   191
C. Carlberg, F. Molnár, *Human Epigenomics*, https://doi.org/10.1007/978-981-10-7614-5_12

12.1   Epigenetics of Hematopoietic Differentiation

Hematopoiesis is the process of lifelong regeneration of blood cells. Every day HSCs give rise to some 10^{11} cells, i.e. hematopoiesis in the bone marrow produces far more cells than any other tissues of the human body. This highly dynamic developmental process includes the self-renewal of multipotent HSCs as well as their differentiation in a hierarchical cascade. Hematopoiesis leads to 11 major lineages of mature blood cells including some 100 phenotypically distinct cell sub-types (Fig. 12.1, *left*).

Most of these cells belong to the immune system, whether innate or adaptive. HSCs differentiate into immature progenitor cells, such as multipotent progenitors (MPPs), which then give rise to the progenitors of the myeloid or lymphoid lineages, called common myeloid progenitors (CMPs) and common lymphoid progenitors (CLPs), respectively. CMPs further differentiate into megakaryocyte-erythrocyte progenitors (MEPs) and granulocyte-monocyte progenitors (GMPs). The cells of myeloid lineage include erythrocytes, platelets and cells of the innate immune system, such as neutrophils, eosinophils and monocytes (which further differentiate into dendritic cells and macrophages). On the other side, the

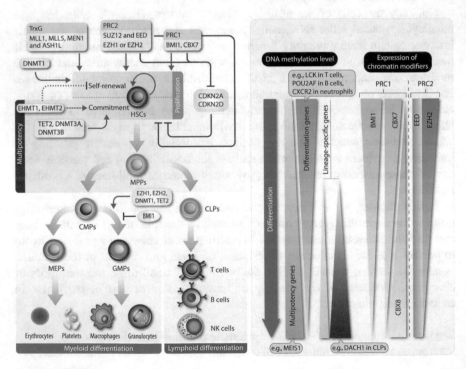

Fig. 12.1 Epigenetics of hematopoiesis. Key chromatin modifiers of HSC self-renewal and their differentiation during hematopoiesis are indicated (*left*). During hematopoietic lineage commitment DNA methylation levels change dynamically (*center*). The expression of key epigenetic regulators changes during hematopoiesis (*right*). More details are provided in the text

lymphoid lineage produces the main cells of the adaptive immune system, B and T cells.

A number of extrinsic and intrinsic factors, such as growth factor stimulated signal transduction pathways, transcription factors and chromatin modifiers, regulate the equilibrium between self-renewal and differentiation of HSCs. In contrast, a disruption or mis-regulation of this process can lead to hematological disorders, such as leukemia, lymphoma and myeloma. Leukemia is categorized into AML (Sect. 7.3), chronic lymphoid leukemia (CLL) and chronic myeloid leukemia (CML). In these diseases the excessive production of white blood cells in the bone marrow significantly raises their level in circulation. Moreover, a disruption in any stage of hematopoiesis affects the production and function of blood cells and can have severe consequences, such as the inability to fight against infections or the risk of uncontrolled bleeding.

Due to their large number and their collection with minimal invasion, cells of the immune system belong to the first that were studied on an epigenome-wide level. Due to the actions of DNMTs and TETs (Sect. 4.1) the genome-wide DNA methylation pattern changes dynamically during hematopoiesis and is very locus specific, i.e. at some genomic regions there is a rise in methylation, while it decreases at other regions (Fig. 12.1, *center*). The latter correlates with the up-regulation of cell-specific genes and their encoded proteins, such as the tyrosine kinase LCK in T cells, the co-factor POU2AF1 in B cells and the chemokine receptor CXCR2 in neutrophils. However, in general the commitment to a specific lineage increases the level of DNA methylation, since a larger set of genes is not anymore needed in these terminally differentiated cells, such as the transcription factor MEIS1, which maintains the undifferentiated state, or the myeloid-specific transcription factor DACH1 in lymphoid cells. Interestingly, the myeloid lineage is the default outcome of hematopoiesis, since their differentiation requires less correction by increased DNA methylation than the lymphoid lineage. Thus, hematopoietic cells can be easily segregated based on their DNA methylation profile.

In parallel with alterations in the DNA methylation during hematopoiesis also key chromatin modifiers are changing their expression. For example, the expression of most members of PcG family changes during HSC differentiation (Fig. 12.1, *right*). The PRC1 components CBX7 and BMI1, which are responsible for the recognition and mono-ubiquitination of H2AK119 (Sect. 6.2), are highly expressed in HSCs but get down-regulated during lineage commitment. In contrast, the CBX7 competitor CBX8 is up-regulated. The PRC2 component EED does not change during hematopoiesis, while the H3K27-KMT EZH2 is down-regulated. Similarly, also members of the TrxG family, such as MLL1 (KMT2A), MLL5 (KMT2E), ASH1L (KMT2H) and MEN1, contribute to hematopoiesis. MLL1 promotes HSC self-renewal and maintains their quiescence. The KMT interacts with the KAT MOF (KAT8) and keeps H4K16Ac levels high at specific genomic loci, so that the genes at these sites stay transcriptionally active. Furthermore, the KMTs EHMT1 and EHMT2 deposit repressive H3K9me2 marks to the epigenome of HSCs (Fig. 12.1, *left*). Accordingly, a mis-regulation of the genes encoding for these chromatin modifiers can result in hematopoietic failure,

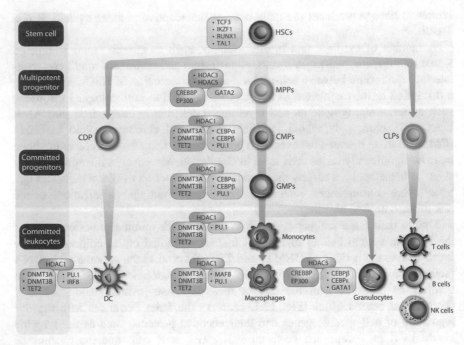

Fig. 12.2 Transcription factors and chromatin modifiers during myeloid differentiation. In each stage of myeloid differentiation (*central vertical axis*) the indicated transcription factors (*yellow boxes*) and chromatin modifiers (*red*, *green* and *blue boxes*) play a key role. In HSCs, the transcription factors IKZF1 and E2A control self-renewal, while the progression into increasingly committed myeloid cells types depends on PU.1 expression. The transcription factor CEBPα cooperates with PU.1 and promotes myeloid identity. Differentiation into macrophages or dendritic cells is controlled by the transcription factors MAFB and IRF8, respectively

such as HSC cell cycle arrest, premature differentiation, apoptosis and defective self-renewal and finally leads to hematological malignancies.

In addition to chromatin modifiers also a number of transcription factors play a role in hematopoiesis. The function of the pioneer transcription factors PU.1, CEBPα and GATA2 had already been discussed in context of transdifferentiation, where, for example, the overexpression of PU.1 and CEBPα can convert fibroblasts into macrophages (Sect. 8.4). In particular in myeloid lineage commitment (Fig. 12.2) the relative expression level of these transcription factors has a central role. PU.1 and CEBPα cooperate in promoting myeloid identity, for example, by interacting with the DNA demethylating enzyme TET2 and directing it to their target genes, so that they get demethylated during hematopoiesis. However, PU.1 can also recruit DNMT3B to its target genes, which then become de novo methylated. This suggests that PU.1 acts as a critical regulator of the methylation status of its target genes and determines, whether these become activated or repressed.

The activity of TET2 may be the key mechanism why myeloid cells are closer to HSCs than lymphoid cells. This fits with the observation that TET2 is mutated in several myeloid malignancies. Moreover, TET2 could link environmental

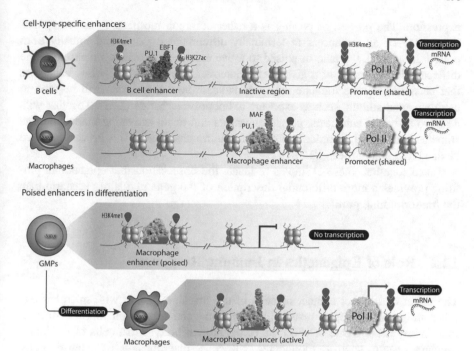

Fig. 12.3 Chromatin-based gene regulation in hematopoietic cells. Due to the use of cell type-specific enhancers and respective transcription factors the same gene can be differently regulated in B cells compared to macrophages (*top*). H3K4me3 marks active TSS regions, while H3K4me1, in combination with H3K27ac, indicates active enhancers. In contrast, in GMPs poised enhancers are marked by H3K4me1 alone, but during differentiation they become active via H3K27 acetylation and the recruitment of the macrophage-specific transcription factor MAF (*bottom*)

conditions, such as nutrient availability (Chap. 13), to myeloid differentiation, since the metabolite 2-hydroxyglutarate inhibits TET2 activity and leads to DNA hyper-methylation.

The fact that alternative cell-specific enhancers and cell-specific transcription factors can regulate a given gene explains the differential expression of genes in hematopoietic cells. For example, the FANTOM5 project (Sect. 2.2) identified some 44,000 enhancers from their large collection of primary human tissues and cells types, i.e. there are approximately two enhancers per protein coding gene. Thus, a given gene can be regulated in B cells via the B cell-specific transcription factor EBF1 binding in combination with PU.1 to a B cell-specific enhancer, while in macrophages the macrophage-specific transcription factor MAF together with PU.1 controls same gene via a macrophage-specific enhancer (Fig. 12.3, *top*).

Furthermore, in multipotent GMP cells an enhancer may be poised, such as being marked only by H3K4me1 but not by H3K27ac, so that the respective gene is not expressed. In contrast, in terminally differentiated macrophages both epigenetic markers are present at the same enhancer and the gene is transcribed (Fig. 12.3, *bottom*). In this way, progenitor cells already prepared a lineage-specific gene for its expression. The process of poising is a rather common mechanism, since approximately 1/3 of all enhancers in terminally differentiated genes had initially been poised. In general, the ratio of poised to active enhancers provides a measure of the differentiation potential of a given cell. Moreover, the amount of poised enhancers that persist in mature immune cells can indicate how these cells will respond to environmental stimuli, such as exposure to lipopolysaccharide (LPS). Together with de novo activated enhancers, poised enhancers can retain activating marks after the stimulus is removed, in order to prepare for future challenges. This process is called "trained immunity" (Sect. 12.2).

Taken together, these examples re-iterate the observation that epigenomic profiling provides a more differential description of the gene regulatory scenario than the transcriptomic profile.

12.2   Role of Epigenetics in Immune Responses

The direct or indirect contact of cells of the immune system with microbes and other antigenic molecules results in effects on gene expression that are stronger than in any other tissue or cell type of the human body. Most cells of the innate immune system, such as monocytes, NK cells, macrophages or dendritic cells, express some 30 different pattern recognition receptors on their surface that respond to the presence of pathogen-associated molecular patterns (PAMPs). PAMPs are molecules, such as LPS on the outer wall of Gram-negative bacteria, that are found in microbes but not in the human host, i.e. they are molecular markers for microbe infection.

The presence of PAMPs causes an inflammatory response that involves significant epigenetic changes in the immune cells. However, the exact response of the immune cells, such as different populations of macrophages, depends on their initial epigenetic state. Moreover, the more than 100 responding genes of the inflammatory cascade differ in their kinetics, i.e. there are fast responding primary target genes, delayed responding secondary targets and late responding tertiary targets. This is reflected by the underlying epigenetic changes in the enhancer and promoter regions of the respective genes of macrophages. The promoter regions of primary target genes typically carry H3K4me3 and H3K27ac marks of active chromatin (Fig. 12.4, *center*). Moreover, these promoters often carry non-methylated CpGs. In contrast, the promoter regions of secondary target genes are first labeled by repressive H3K27me3 marks. The removal of H3K27me3 marks and CpG demethylation from these promoter regions after PAMP exposure and the introduction of H3K4me3 and H3K27ac marks takes time. This explains the delayed response in the expression of the respective genes.

The stimulation of macrophages with PAMPs also changes the chromatin status at enhancer regions, such as the de novo deposition of H3K4me1 and H3K27ac marks (Fig. 12.4 *left*). The activation of some of these enhancers is only transient, while others are marked more persistently and retain in this way a "memory" of

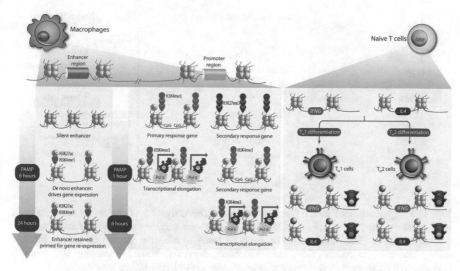

Fig. 12.4 Epigenetic modifications in immune cells. Enhancer (*left*) and promoter regions (*center*) of macrophages show a differential response after the exposure with PAMPs, such as LPS. Primary responses (after 1 hour) and secondary responses (after 6 hours) are distinguished. Even after removal of PAMP stimulation some enhancer regions can retain their activation status for 24 hours or longer. This memory effect is part of trained immunity. Differential epigenetic programing of the regulatory regions of key cytokine genes, such as *IFNG* and *IL4*, is the key event in the polarization of T_H cell subtypes (*right*). In T_H1 cells the *IFNG* gene carries marks of active chromatin (*green*) and the gene is induced after antigen exposure. In contrast, in the same cells the *IL4* locus has repressive histone markers (*red*) and stays repressed. In T_H2 cells the reverse process applies, i.e. the *IL4* gene is induced and the *IFN* gene stays repressed

the PAMP exposure, i.e. of the contact with microbes. This epigenetic programing phenomenon of trained immunity applies also to other cells of the innate immune system, such as monocytes and NK cells (Fig. 12.5). It is mediated by histone modifications, DNA methylation and the actions of miRNAs and long ncRNAs. The rather long half-life of the latter molecules makes them well suited for a persistent programing of the epigenome.

Trained immunity enables innate immune cells to react with a quantitatively different response, i.e. a higher magnitude of gene expression when they are re-challenged with a pathogen (Fig. 12.6 *top left*). This response can in part also be qualitatively different, such as via the expression of an alternative pattern recognition receptor (Fig. 12.6 *bottom left*). A key mechanism in trained immunity is enhancer poising, i.e. the addition of persistent histone marks, such as H3K4me1, enabling a strong response after restimulation (Fig. 12.6 *right*). Interestingly, most immune cells are very mobile and throughout the human body experience many different microenvironments. This leads to wide range of different signals and respective adaptive reprograming of the cells. Accordingly, the transfer of immune cells, such as macrophages, from one tissue to another results in an extensive reprograming of their enhancer repertoire. In general, there is equilibrium between

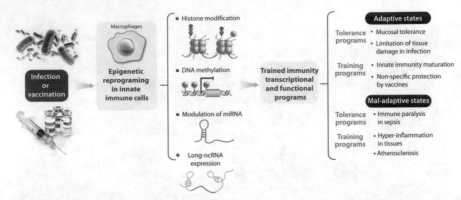

Fig. 12.5 Epigenetics of innate immune response. The activation of the innate immune cells, such as monocytes, macrophages or NK cells, leads to their epigenetic reprograming, known as trained immunity. This innate immune memory leads to adaptive states that protect the host during and after infections. In certain situations, however, trained immunity can result in mal-adaptive states such as immune paralysis after sepsis or hyper-inflammation

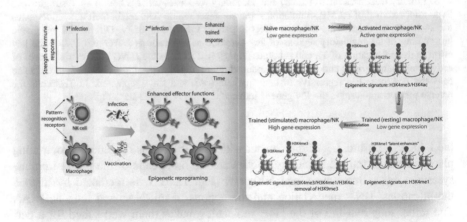

Fig. 12.6 Trained immunity. Enhanced inflammatory and anti-microbial properties of innate immune cells (*top left*) are a memory phenomenon that is referred to as trained immunity. It is based on epigenetic reprograming of innate immune cells, such as macrophages and NK cells, and results, for example, in the increased and/or alternative expression of genes encoding pattern recognition receptors (*bottom left*). The first round of stimulation of the cells leaves persistent H3K4me1 marks on enhancer regions. This enhancer poising enables them to respond faster and stronger to a restimulation (*right*)

the persistence of an epigenome instructed by previous stimuli and the reprograming in response to a changing environment.

The cells of the adaptive immune system, T and B cells, use highly specific receptors on their surface, such as T cell receptors and B cell receptors, in order and to recognize antigens derived from pathogens. Antigen binding to these receptors activates signal transduction pathways that potently trigger the expression of cytokine genes. Also in this case specific epigenetic changes to enhancer and promoter regions are involved. For example, the differentiation of T helper (T_H) cells into T_H1 and T_H2 subtypes involves epigenetic programing that primes these cells to increase after antigen exposure either the expression of the genes encoding for the cytokines interferon γ (*INFG*) or interleukin 4 (*IL4*), respectively (Fig. 12.4 *right*). While the T_H1 response results in antigen clearance, the T_H2 response is typical for allergic reactions, i.e. the differential epigenetic programing of these cells causes a clearly different physiological response. The epigenetic changes involve histone modifications but also the appearance of 5hmC marks at promoter regions and demethylation via TET2. The latter modifications are very stable and can last over 20 and more replication cycles of long-lived memory T cells.

12.3 Epigenetics of Immune Diseases

Inappropriate activation of the immune system can lead to a number of diseases, such as the autoimmune disease multiple sclerosis (MS) or the allergic reactions of the respiratory tract in asthma. Immune responses often vary in the balance of pro-inflammatory and anti-inflammatory cytokines, i.e. in the amount of possible collateral tissue damage. Thus, gene expression patterns and the underlying epigenetic programing of immune cells are key factors in immune-mediated diseases. Interestingly, individuals with autoimmune and/or inflammatory diseases have a clearly different epigenetic profile than healthy controls. For example, altered DNA methylation patterns of immune cells occur in multiple immune-mediated diseases, such as MS, systemic lupus erythematosus (SLE) or Crohn disease. Patients with MS, in comparison to healthy subjects, show lower levels of 5hmC marks in their immune cells due to low expression of TET2. In contrast, SLE patients often have elevated 5hmC levels due to increased expression of TET2 and TET3 in T_H cells. This parallels with low global H3 and H4 acetylation and high H3K9 methylation levels in these cells.

Genome-wide profiling of immune cell subsets can help to identify the epigenomic basis of diseases, such as asthma. For example, when comparing healthy controls with asthmatic patients, a marker for active and poised enhancers, H3K4me2, is significantly increased in T_H2 cells of patients (Fig. 12.7). The low amount of available primary cells from patients requires the use of highly sensitive versions of ChIP-seq assay.

The cell-specificity of enhancers (Sect. 12.2) highlights these genomic regions as key sites for detecting disease-causing epigenetic variants. The genetic risk for asthma was previously discovered on a population level by GWAS (Sect. 2.3). Most of these SNPs are located outside of protein-coding regions but may disrupt or generate transcription factor binding sites within enhancer regions. EWAS was already successfully applied for the mapping of QTLs associated with gene expression (Sect. 9.2). However, only epigenomic assays, such as ChIP-seq, provide an

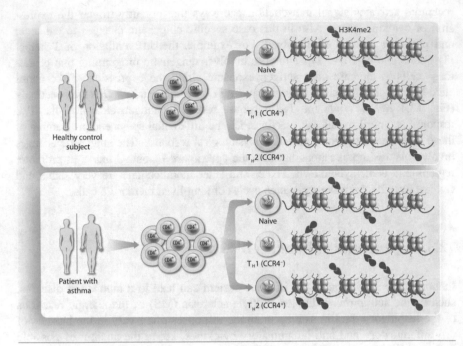

Fig. 12.7 Epigenomic profiling of T cells in asthma. PBMCs of healthy control subjects (*top*) and patients with asthma (*bottom*) were sorted via the chemokine receptor CCR4 into naïve T cells, T_H1 and T_H2 cells. Profiling for active and poised enhancers was done via H3K4me2 ChIP-seq. Main differences between patients and controls are seen for T_H2 cells

experimental proof that these non-coding SNPs have a functional consequence. Moreover, asthma is a master example of a disease in which environmental exposure with natural and synthetic compounds causes dynamic changes of the epigenome. Thus, epigenetic profiling of cell subsets for various epigenetic markers, such as accessible chromatin, DNA methylation, histone modifications, transcription factor binding and chromatin modifier association, is an important tool for a molecular understanding of disease and can provide hints for a possible therapy.

Antigens stimulate the clonal expansion of naïve cytolytic T cells, i.e. the growth of antigen-recognizing subsets. This is followed by cellular differentiation, which activates a network of effector genes, such as those encoding for cytokines, via the demethylation of their genomic regions (Fig. 12.8). The resulting effector T cells are able to eliminate directly and indirectly antigen-presenting cell. Anergy is an epigenetically induced dys-functional state of T cells that often occurs during prolonged exposure to an antigen, such as in cancer and in chronic infections. This results in de novo methylation of effector genes and their inactivation, i.e. the exhausted T cells show no further immune response. The methylation of the effector genes can be blocked and/or reversed by the application of a demethylating agent, such as such as decitabine (Sect. 4.1), i.e. the exhausted T cells get

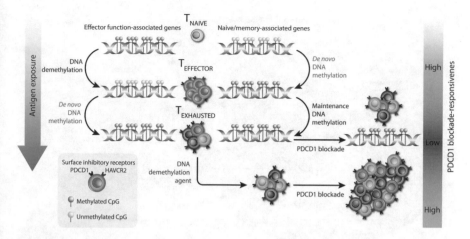

Fig. 12.8 Epigenetic reprograming of T cells. In naïve cytolytic T cells (CD8⁺), effector-function associated genes are methylated and inactive, but become demethylated when the T cells, after antigen exposure, become effector cells after antigen exposure (*left*). However, the persistent exposure to antigen leads to T cell exhaustion and de novo methylation of these genes. In contrast, genes associated with the function of naïve or memory cells are initially unmethylated, get de novo methylated after antigen exposure and keep this status after exhaustion of the cells (*right*). A PDCD1 blockage can only effectively rejuvenate the exhausted T cells, when they first have been treated with a DNA demethylating agent (*bottom*)

epigenetically reprogramed. In this case, the blockage of the surface inhibitory receptor PDCD1 (also called PD1) is far more effective, i.e. the T cells start again their growth and effector function. The combination with an epigenetic mediator gives immune checkpoint blockage of T cells a wider approach for treating both cancer and chronic infections. For example, Hodgkin lymphoma patients, who received a DNMT inhibitor before a treatment with immune checkpoint inhibitors showed a higher rate of complete remission in their cancer.

The general use of inhibitors of chromatin modifiers in cancer, such as KDAC and DMNT inhibitors, in cancer was already discussed in Sect. 10.5. More specifically, Fig. 12.8 showed an example how these small-molecule inhibitors enhance the efficacy of immunotherapeutic agents, such as a blockage of the interaction between PDCD1 on the surface of cytolytic T cells and CD274 on cancer cells. Fig. 12.9 provides an overview how these epigenetic inhibitors can boost the anti-tumor immune response of the host. For example, the DNMT inhibitors azacitidine and decitabine induce the expression of *HLA* genes encoding for major histocompatibility complexes (MHCs) and of tumor antigens, such melanoma-associated antigen 1 (MAGEA1). This increases the visibility of the cancer cell to cytolytic T cells and their subsequent elimination. In addition, decitabine increases the sensitivity of cancer cells to growth inhibition by type 1 interferons. This results in a "viral mimicry," in which DNA demethylation activates the transcription of endogenous retroviral elements in the cancer cells leading to a dsRNA-mediated immune response.

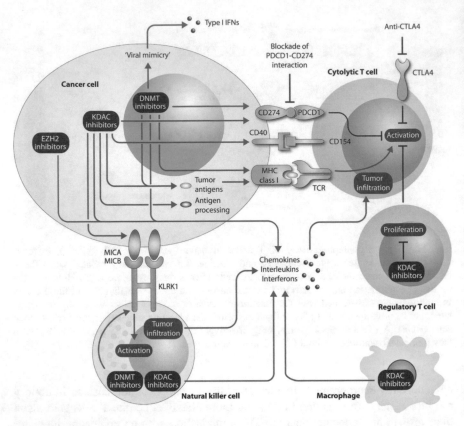

Fig. 12.9 Inhibitors of chromatin modifiers in immunotherapy. Epigenetic inhibitors also may play an important role in immuno-oncology. The role of DNMT inhibitors, KDAC inhibitors and EZH2 inhibitors both in cancer cells and in different types of immune cells is discussed in the text

KDAC inhibitors also modulate the expression of MHC proteins, co-stimulatory CD40 molecules and tumor antigens. Moreover, they affect the antigen-processing machinery and change in chemokine expression in both cancer and immune cells. Furthermore, they suppress T_H cells and induce of the NK cell receptor ligands MHC class I polypeptide-related sequence (MIC) A and B. Inhibitors of the PRC2 component EZH2 (KMT6A) increase the expression of chemokines CXCL9 and CXCL10 attracting T cells and improving tumor clearance.

Key Concepts
- The self-renewal and differentiation of HSCs is regulated by a number of extrinsic and intrinsic factors, such as signal transduction pathways stimulated by growth factors, as well as transcription factors and chromatin modifiers.

(continued)

Key Concepts (continued)
- Most of the some 100 different blood cell types belong to the immune system and differ in their epigenetic programing, in particular at cell-specific enhancer regions.
- Hematopoietic cells can be easily segregated based on their DNA methylation profile.
- A mis-regulation of the genes encoding for chromatin modifiers can result in hematopoietic failure, such as HSC cell cycle arrest, premature differentiation, apoptosis and defective self-renewal, and finally leads to hematological malignancies.
- TET2 links environmental conditions, such as nutrient availability, to myeloid differentiation, since the metabolite 2-hydroxyglutarate inhibits TET2 activity and leads to DNA hyper-methylation.
- Trained immunity is an epigenetic process where de novo activated and poised enhancers retain activating marks after the stimulus is removed, in order to prepare for future challenges.
- Immune cells have a memory function that is based on epigenetic changes, such as DNA methylation and histone modifications, and monitors the close relationship between immune challenges and the effects on chromatin.
- A key mechanism in trained immunity is enhancer poising, i.e. the addition of persistent histone marks, such as H3K4mc1, enabling a strong response after restimulation.
- Immune-mediated diseases, such as autoimmune diseases and inflammation, are clinically heterogeneous, but all develop from the interplay of genetic susceptibility and environmental or lifestyle factors, i.e. the balance between genetics and epigenetics.
- Individuals with autoimmune and/or inflammatory diseases have a clearly different epigenetic profile than healthy controls.
- Epigenetic profiling of cell subsets for various epigenetic markers, such as accessible chromatin, DNA methylation, histone modifications, transcription factor binding and chromatin modifier association, is an important tool for a molecular understanding of disease and can provide hints for a possible therapy.

Additional Reading

Alvarez-Errico D, Vento-Tormo R, Sieweke M et al (2015) Epigenetic control of myeloid cell differentiation, identity and function. Nat Rev Immunol 15:7–17

Avgustinova A, Benitah SA (2016) Epigenetic control of adult stem cell function. Nat Rev Mol Cell Biol 17:643–658

Ghoneim HE, Fan Y, Moustaki A et al (2017) De novo epigenetic programs inhibit PD-1 blocka-demediated T cell rejuvenation. Cell 170:142–157

Hu D, Shilatifard A (2016) Epigenetics of hematopoiesis and hematological malignancies. Genes Dev 30:2021–2041

Netea MG, Joosten LA, Latz E et al (2016) Trained immunity: a program of innate immune memory in health and disease. Science 352:aaf1098

Tough DF, Tak PP, Tarakhovsky A et al (2016) Epigenetic drug discovery: breaking through the immune barrier. Nat Rev Drug Discov 15:835–853

Vahedi G, Richard AC, O'Shea JJ (2014) Enhancing the understanding of asthma. Nat Immunol 15:701–703

Winter DR, Jung S, Amit I (2015) Making the case for chromatin profiling: a new tool to investi-gate the immune-regulatory landscape. Nat Rev Immunol 15:585–594

Chapter 13
Nutritional Epigenomics

Abstract The energy status of tissues and cell types is the most important information for the human body, in order to interpret and integrate environmental conditions. Metabolic pathways communicate with chromatin and provide information about nutrient availability and energy status via the levels of key metabolites, such as AMP, NAD^+, SAM and acetyl-CoA. These metabolites act as co-factors and substrates of chromatin modifiers, such as KMTs, KATs, KDACs and TET enzymes. Accordingly, gene expression programs that control cell fate decisions, such as proliferation, differentiation and autophagy, are modulated by the metabolic status. The results of these epigenetic events may be memorized in the epigenome of metabolic organs. This means that the lifestyle of individuals, i.e. primarily their daily diet, creates a metabolic memory. Epigenetic programing during embryogenesis but also in adult life may explain the missing genetic heritability of the susceptibility for complex diseases, such as T2D.

In this chapter, we will define different epigenetic mechanisms that process information provided by dietary molecules. We will learn that many chromatin-modifies are susceptible to changes in the levels of intermediary metabolites and respond to changes in nutrient intake and metabolism. Finally, will get insight into the concepts of epigenetic programing of metabolic tissues.

Keywords Nutrigenomics · acetyl-CoA · NAD^+ · SAM · folate · one-carbon metabolism · muscle differentiation · adipogenesis · T2D

13.1 Epigenetic Mechanisms Linking Metabolites and Gene Transcription

The availability of energy substrates from nutrition is essential for all tissues and cell types of the human body. Accordingly, this important environmental input results in the activation of signal transduction pathways within cells. These pathways stimulate gene expression programs that integrate the information about the nutritional status with the goal to preserve cellular homeostasis. Chromatin modifiers are important nodes in these signal transduction and integration processes,

© Springer Nature Singapore Pte Ltd. 2018

C. Carlberg, F. Molnár, *Human Epigenomics*, https://doi.org/10.1007/978-981-10-7614-5_13

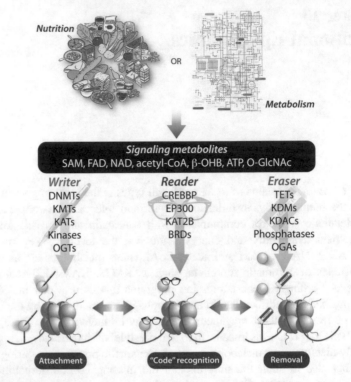

Fig. 13.1 Metabolites and transcriptional regulation are linked by epigenetics. Changes in nutrition or fluctuations in metabolism affect the transcriptional responses of metabolic tissues. A number of intermediary metabolites are co-substrates and/or co-factors of chromatin modifiers, i.e. these enzymes act as metabolic sensors. Writer enzymes create covalent chromatin marks, reader enzymes recognize these marks and eraser enzymes remove them (Sect. 5.1). This results in changes of local chromatin structure and has consequences for the activity and regulation of the neighboring genes

since many of them use intermediary metabolites, such as acetyl-CoA, α-KG, NAD$^+$, FAD, ATP or SAM, as co-substrates and/or co-factors (Fig. 13.1).

The activity of most chromatin modifiers critically depends on intra-cellular levels of essential metabolites. For example, TET enzymes require oxygen, Fe(II) and α-KG for the catalysis of 5mC oxidations (Sect. 8.4). Thus, chromatin modifiers can act as sensors of the metabolic status of a cell and translate the metabolic information into dynamic post-translational histone modifications that coordinate adaptive transcriptional responses. The activities of the respective gene regulatory networks control cell fate decisions.

For example, the NAD$^+$/NADH ratio reflects the cellular redox state and is inversely proportional to the energy state of the cell. During fasting, i.e. at low levels of nutritional metabolites, the intra-cellular concentration of NAD$^+$ raises. This leads to an increase in sirtuin activity and the deacetylation of their target proteins (Fig. 13.2 *left*). The latter are histone proteins, but also transcription factors or their co-factors, such as p53 and peroxisome proliferator-activated receptor

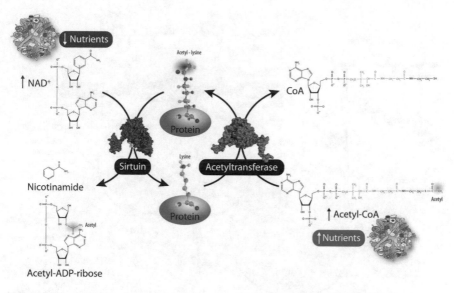

Fig. 13.2 The relation of protein acetylation and cellular metabolism. NAD$^+$ (*left*) acts as a co-factor for KDACs of the sirtuin family that deacetylate proteins, which had been acetylated by KATs using acetyl-CoA (*right*). Thus, the acetylation status of key regulatory proteins reflects the cellular concentration of NAD$^+$ and acetyl-CoA, i.e. of low or high nutritional status, respectively

gamma, co-activator 1 alpha (PPARGC1A). Accordingly, the beneficial effects of calorie restriction on metabolic health are based on NAD$^+$-stimulated sirtuins. Since the NAD$^+$ concentrations fluctuate in a circadian manner, sirtuin-mediated epigenetic mechanisms are linked to the molecular clock of transcriptional regulation. Moreover, the metabolite D-β-hydroxybutyrate (βOHB) acts as an inhibitor of the KDACs HDAC1, HDAC2 and HDAC3. During fasting or calorie restriction, levels of βOHB increase and promote histone acetylation.

In contrast, in the feeding state ingested nutrients enter the catabolic pathways of intermediary metabolism and acetyl-CoA is produced. Augmented acetyl-CoA concentrations stimulate KAT activity, so that their target proteins get acetylated (Fig. 13.2 *right*). When the target proteins are histones, the acetylation of chromatin leads to open chromatin. This stimulates the expression of genes involved in metabolic processes, such as lipogenesis and adipocyte differentiation. Moreover, at hypoxic conditions the transcription factors HIF1α and HIF1β recruit the KATs EP300 and CREBBP resulting in increased histone acetylation at the chromatin loci of their binding sites. Oxygen levels also directly regulate the expression and catalytic activity of KDMs, such as KDM5C, since these enzymes are 2-oxo-glutarate-dependent dioxygenases (Sect. 8.4).

The methyl donor substrate SAM connects the processes intermediary metabolism and DNA methylation. After the methyl group of SAM is transferred to a histone or to genomic DNA (Fig. 13.3), the product S-adenosylhomocysteine (SAH) is recycled back to SAM. Interestingly, SAH is a negative feedback regulator of KMTs, i.e. the SAM/SAH ratio (also referred to as the "methylation index") is critical for histone and DNA methylation. This is another example how variations

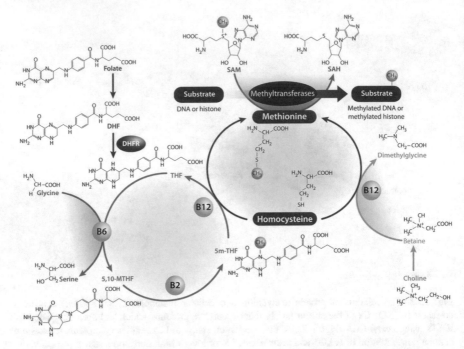

Fig. 13.3 Link of methyltransferases with one-carbon metabolism. DNMTs and KMTs use SAM as a methyl group donor for the methylation of genomic DNA and histones, respectively. The generated SAH is converted in multiple steps back to homocysteine (details not shown). Vitamin B12-dependent reactions utilize carbons derived from either choline or folate, in order to convert homocysteine back to methionine. Steps that require the vitamins B6 and B2 are also shown, but for simplicity, the respective converting enzymes are not indicated. More details are provided in the text

in the cellular metabolic state affect the relative concentration of enzymatic co-factors and influence the activity of chromatin modifiers.

A derivative of the B vitamin folate, tetrahydrofolate, feeds the cyclic one-carbon pathway (Fig. 13.3) by serving as a methyl group donor. This demonstrates the direct connection between nutrition and epigenetics. Thus, methyl donors are critical for epigenetic programing during embryogenesis. A high homocysteine level is an established biomarker for the disturbance of the one-carbon metabolism and related to low concentrations of folate, vitamins B12 and B6, choline and betaine. This may cause an elevated risk of pre-mature delivery, low birth weight and neural tube defects. Moreover, a low dietary intake of folate or methionine increases the risk of colon adenomas, while *in utero* exposure to higher folate is associated with a reduced risk of childhood acute lymphoblastic leukemia, brain tumors and neuroblastoma.

The enzyme methylenetetrahydrofolate reductase, which is encoded by the gene *MTHFR*, catalyzes the conversion of 5,10-methylenetetrahydrofolate to 5-methyltetrahydrofolate (Fig. 13.3). 10–15% of Europeans carry a C677T missense

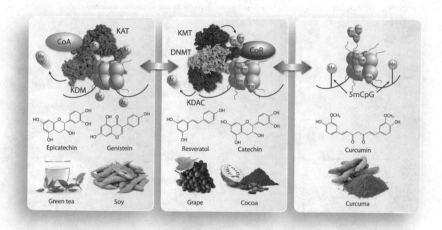

Fig. 13.4 Natural compounds regulate the activity of chromatin modifiers. Plant-origin natural compounds modulate the activity of chromatin modifiers, such as KATs and KDMs in open chromatin (*left*), KMTs, DNMTs and KDACs in facultative heterochromatin (*center*) and methylated CpGs in heterochromatin (*right*). In this way they affect the epigenetic status of most human tissues and cell types

SNP (rs1801133) on both alleles, which reduces the activity of the enzyme by some 50%. Accordingly, individuals with a TT genotype are affected more by a low folate intake than those with the CC or CT allele.

Finally, a wide spectrum of secondary metabolites that humans take up from fruits, vegetables, teas, spices and traditional medicinal herbs, such as genistein, resveratrol, curcumin and polyphenols from green tea, coffee and cocoa, are able to modulate the activity of chromatin modifiers (Fig. 13.4).

13.2 Cell Decisions via the Metabolome-Epigenome Axis

There are a number of mechanisms how metabolites trigger the epigenome (Fig. 13.5). For example, α-KG is an essential co-substrate of Jumonji domain-containing KDMs (Sect. 6.3) and TET enzymes (Sect. 13.1). Metabolites that are structurally similar to α-KG, such as 2-hydroxyglutarate, succinate and fumarate, can block the α-KG binding site and inhibit these enzymes. Under hypoxic conditions (S)-2-hydroxyglutarate is produced by the enzyme lactate dehydrogenase (LDH), while the isomer (R)-2-hydroxyglutarate is created in cancer cells by mutant forms of the enzymes IDH1 and IDH2 (Fig. 13.5, *left*, Sect. 10.1). In this way, comparably different conditions, such as hypoxia and gene mutations, can lead to the production of a metabolite that causes hyper-methylation of both

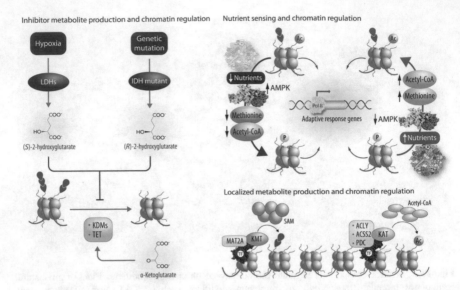

Fig. 13.5 Triggering the epigenome by metabolites. In hypoxia LDHs produce R-2-hydroxyglutarate, while mutant IDH leads to the accumulation of S-2-hydroxyglutarate (*left*). Both isomers compete with α-KG for binding to TET enzymes and Jumonji domain-containing KDMs leading to DNA and histone methylation. Low nutrient availability, such as low levels of methionine and acetyl-CoA, leads to changes in histone modification patterns, such as AMPK-mediated phosphorylation, and respective adaptive gene expression (*top right*). This scenario reverses at high nutrient levels. The direct recruitment of metabolic enzymes, such as MAT2A, ACYL, ACSS2 and PDC, to specific chromatin loci allows the local production of SAM and acetyl-CoA, respectively, and accordingly KMT- and KAT-mediated chromatin methylation and acetylation (*bottom right*)

histone proteins and genomic DNA. Similarly, the accumulation of succinate or fumarate in tumors that are deficient for the enzymes succinate dehydrogenase or fumarate hydratase can lead to the inhibition of α-KG-dependent enzymes and also causes hyper-methylation.

Some nutrient sensing mechanisms were already discussed in Sect. 13.1. Another example of metabolite sensing is that of the enzyme AMPK (AMP-activated protein kinase), which is controlled in its activity by the AMP/ADP ratio. When cells consume more ATP than they are producing, i.e. at conditions of low nutrient availability, the AMP concentration is raising as a signal of energetic stress. AMP binds to the γ-subunit of the AMPK heterotrimer and activates the kinase. Since histones belong to the AMPK substrates, the low energy status of the cell is marked via histone phosphorylation (Fig. 13.5, *top right*). Thus, insults to the energy status of the cell are memorized on the level of histone modifications and can be translated into functional outputs via adaptive gene regulation. In contrast, a high nutritional level results in low AMP levels, no AMPK activity, a modified histone phosphorylation pattern and the activity of a different set of genes.

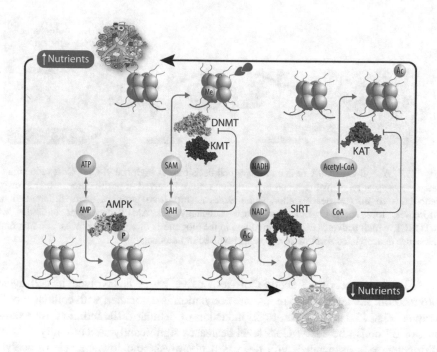

Fig. 13.6 Epigenetic sensing of the nutritional state. A high nutritional state of a cell (*top*) is represented by the abundance of the metabolites ATP, SAM, NADH and acetyl-CoA, while in case of low nutrient levels (*bottom*) the metabolites AMP, SAH, NAD$^+$ and CoA are predominant. Accordingly, at high nutrient concentrations, KMTs and KATs are stimulated, while at low concentrations AMPK and KDACs of the sirtuin family are activated and DNMTs and KATs are repressed. This results in histone methylation and acetylation or histone phosphorylation and deacetylation, respectively

The enzyme MAT2A (methionine adenosyltransferase 2A) converts some 50% of the daily intake of methionine into SAM. Interestingly, the transcription factor MAFK recruits the enzyme together with KMTs to specific chromatin sites, which then locally get methylated (Fig. 13.5, *bottom right*). Similarly, the acetyl-CoA producing enzymes ACLY (ATP-citrate lyase), ACSS2 (acyl-CoA synthetase short-chain family member 2) and PDC (pyruvate dehydrogenase complex) can also be recruited to specific genomic sites. At these chromatin loci they generate acetyl-CoA, which is used by KATs for histone acetylation.

The metabolic state of a cell can be expressed by the ATP/AMP ratio, the SAM/SAH ratio, the NADH/NAD$^+$ ratio and the acetyl-CoA/CoA ratio (Fig. 13.6). Under high nutrient concentrations, such as abundant availability of methionine and glucose, SAM activates KMTs and acetyl-CoA stimulates KATs, thus leading to histone methylation and acetylation, respectively. In contrast, at low nutrient levels, such as during fasting, AMP activates AMPK and NAD$^+$ stimulates sirtuins (Sect. 6.3) resulting in histone phosphorylation and deacetylation. Moreover, in parallel SAH inhibits DNMTs and CoA blocks KATs.

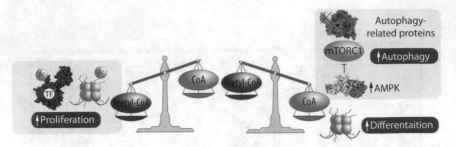

Fig. 13.7 Acetyl-CoA/CoA ratio and major cell decisions. A high acetyl-CoA/CoA ratio stimulates the acetylation of histones and transcription factors involved in proliferation, while a lower ratio leads to histone deacetylation and induces a shift from proliferation to differentiation. Moreover, low acetyl-CoA levels induce autophagy via AMPK-dependent inhibition of mTORC1, which activates enzymes involved in the biogenesis of autophagosomes, and favoring the active deacetylated form of key proteins involved in autophagy

A large proportion of acetyl-CoA-responsive genes is involved in cell cycle progression, i.e. an increase in histone acetylation is associated with cellular proliferation (Fig. 13.7). However, upon induction of cellular differentiation, for example, of ES cells, the acetyl-CoA level decreases significantly. Accordingly, loss of pluripotency is associated with decreased glycolysis and lower levels of acetyl-CoA and histone deacetylation. Moreover, the acetyl-CoA level also affects cell survival and death decisions. For example, a low acetyl-CoA level induces the catabolic process of autophagy, which is crucial for organelle quality control and cell survival during metabolic stress (Fig. 13.7). Thus, the acetyl-CoA/CoA ratio is an important regulator of major cellular decisions.

13.3   Epigenetic Regulation of Metabolic Tissue Differentiation

Skeletal muscle and adipose tissue are metabolic organs that constitute more than half of the human body mass. Their relative amount is very variable and depends on environmental factors, such as physical activity and nutritional intake.

Stem cells in adult muscle, referred to as satellite cells, remain quiescent until they are activated by environmental signals, such as those following injury or stress. Like in HSCs (Sect. 12.1), symmetric and asymmetric cell divisions balance self-renewal of satellite cells and their differentiation to muscle cells. For satellite cell self-renewal the expression of the master transcription PAX7 is sufficient (Fig. 13.8). However, this self-renewal process is inhibited by the PRC1 component BMI1, which deposits H2AK119Ub1 marks (Sect. 12.1) at specific chromatin loci, such as those of the cycle inhibitor CDKN2A, while the PRC2 components EZH1 and EZH2 place H3K27 methylation marks (Sect. 6.2) at the same and other critical genomic regions. In addition, the arginine methyltransferase PRMT5 (Sect. 5.1) delivers H3R8me2 marks at the locus of the *CDKN1A* gene. Moreover,

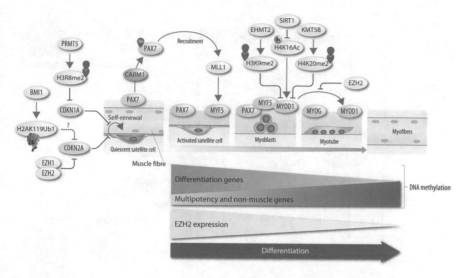

Fig. 13.8 Epigenetic regulation of muscle differentiation. The stem cells in adult muscles remain quiescent until their function is required, such as during muscle regeneration. Master transcription factors drive the progression through the myogenic differentiation program. Epigenetic changes during stem cell self-renewal and differentiation are indicated. More details are provided in the text

the expression of transcription factor MYF5 (myogenic factor 5) is activated and drives the progression of the cells through the myogenic differentiation program.

During myogenic differentiation, DNA methylation at pluripotency and non-muscle genes increases, while it decreases at loci related to muscle specification, such as that of the *MYF5* gene (Fig. 13.8). In addition, the arginine methyltransferase CARM1 methylates PAX7, which then recruits the H3K4 KMT MLL1, a TrxG component, to its target genes, such as *MYF5*. The progression of differentiation of myoblasts into myotubes requires the expression of the transcription factor MYOD1 (myogenic differentiation 1). *MYOD1* expression depends on the metabolic state of the myoblasts and is controlled via the NAD⁺-sensitive KDAC SIRT1 leading to an increase in H4K16 acetylation at the *MYOD1* locus.

Mesenchymal stem cells (MSCs) are able to differentiate into multiple meso-dermal cell lineages, such as adipocytes and osteoblasts (Fig. 13.9). During aging, the differentiation equilibrium of MSCs is dys-regulated, so that adipogenesis is favored over osteogenesis. The fatty acid-sensitive transcription factor PPARγ (peroxisome proliferator-activated receptor γ) is a master regulator of adipogenesis in MSCs. In contrast, signal transduction pathways that are induced by the peptide WNT10A inhibit adipogenesis and promote osteogenesis.

EZH2-mediated H3K27me3 chromatin marking is an important epigenetic control point of adipogenesis (Fig. 13.9). In contrast, KDM6A and KDM6B remove H3K27me3 marks and in this way de-repress in this way key transcription factors involved in osteogenic differentiation. Furthermore, the KMT EHMT2 places

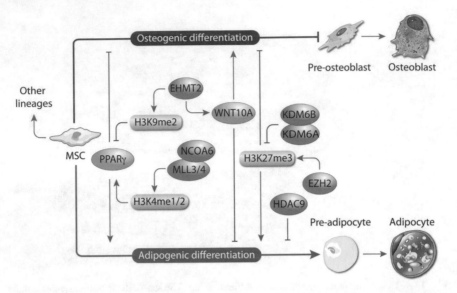

Fig. 13.9 Epigenetic regulation of MSC differentiation. Under the critical control of the transcription factor PPARγ and the signaling peptide WNT10A MSCs differentiate to adipocytes and osteoblasts, respectively. More details are provided in the text

H3K9me2 marks that prevent *PPARG* expression and stimulate WNT10A expression. This prevents the adipogenic differentiation and promotes the formation of osteoblasts. Furthermore, the H3K4 KMTs MLL3 and MLL4 form a complex with the co-activator NCOA6 (nuclear receptor co-activator 6), interact with PPARγ and induce the transcription of adipogenic genes.

13.4 Epigenomics of Intergenerational Metabolic Disease

The concept of epigenetic programing via nutritional compounds suggests that dietary interventions of human adults, such as calorie restriction, "Mediterranean" or "Nordic" diet, all can affect the chromatin status of the individuals and lead to the expression of genes being beneficial to metabolic health. This implies that epigenetic states, which are initially fixed during embryogenesis, may shift in response to intrinsic and environmental factors, such as nutritional compounds. The epigenetic drift of metastable epialleles was already discussed in the context of epigenetic memory and transgenerational inheritance (Sect. 9.1). For example, changes in diet or in metabolism being associated with obesity can cause an epigenetic drift that may be inherited to the following generations. If this concept holds true, the worldwide growing epidemic of obesity and metabolic disease may lead to a significant epigenetic predisposition for the metabolic syndrome in the subsequent generations resulting in a vicious cycle. Moreover, epigenetic modifications acquired during chronological aging reduce the capacity for homeostatic responses to nutritional stress, such as overnutrition.

In agouti viable yellow (A^{vy}) mice, the *Asip* gene is under the control of the DNA methylation-sensitive retrotransposon IAP (Sect. 9.1). This represents an interesting model for the testing of nutrition-triggered transgenerational inheritance. Female wild-type a/a mice are either supplemented or not with methyl donors, such as folate, vitamin B12 and betaine, as early as two weeks before mating with male A^{vy}/a mice as well as during pregnancy and lactation. The F1 generation of non-supplemented mothers displays the expected number of A^{vy} yellow fur color phenotypes, but the offspring of supplemented mothers shifts toward the wild-type brown coat color phenotype, i.e. the yellow A^{vy} phenotype is repressed via methylation of the IAP retrotransposon.

The methylation pattern of the IAP element is established early in embryogenesis. The inheritance of an epigenetic programing to the next generation indicates that at least metastable epialleles, such as IAP, resist the global demethylation of the genome before pre-implantation (Sect. 8.1). Interestingly, when A^{vy} mice are fed with a soy polyphenol diet, which causes changes in their DNA methylation patterns, their offspring is protected against diabetes, obesity and cancer across multiple generations.

In contrast, in a mouse model of maternal undernutrition, which causes low birth weight and glucose intolerance in the offspring, the exposure to suboptimal nutrition during fetal development leads to changes in the germ-cell DNA methylome of male offspring, even when these males were nourished normally after weaning. These phenotypic differences are transmitted through the paternal line to the F2 offspring. In the genome-wide view of DNA methylation, more than 100 genomic regions in the F1 sperm from maternally undernourished male offspring were hypo-methylated compared with the equivalent regions of sperm from control offspring. In this model the presence of novel hypo-methylated regions suggests that PGCs from nutritionally restricted fetuses do not completely re-methylate their DNA.

Also in humans critical disturbances in energy metabolism can lead to stable epigenetic changes that are maintained through the germ line and may affect the health of the next generations (Sect. 9.2). For example, alterations in paternal diet, such as high-fat or low-protein diets, or intra-uterine exposure to maternal caloric restriction (*Dutch Hunger Winter*, Sect. 8.4) can result in increased T2D risk in offspring (Fig. 13.10). Moreover, the DNA methylome of pancreatic islets from individuals with T2D and that from healthy individuals differs at more than 800 genomic loci. Some of these methylation changes correlate with mRNA levels suggesting differences in β cell function.

A number of nutrigenomic approaches aim to maintain wellbeing, promote health and open up new therapeutic strategies, such as a possible reprograming of the epigenome of metabolic organs through personalized diet including natural compounds that modulates the activity of chromatin modifiers and transcription factors. This concept suggests that lifestyle changes, such as increased physical activity and consecutively weight loss can have a beneficial effect on the epigenome and thus lowering the risk for suffering from the metabolic syndrome.

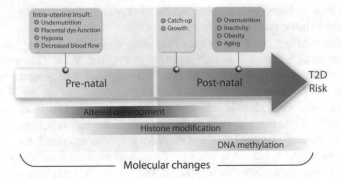

Fig. 13.10 Pre-natal programing of diabetes risk. Intra-uterine stressors, such as maternal undernutrition or placental dys-function, can lead to impaired blood flow, nutrient transport or hypoxia. These events initiate abnormal patterns of development, histone modification and DNA methylation that increase the T2D risk in adult life. In addition, post-natal environmental factors, such as accelerated growth, overnutrition, obesity, inactivity and aging, lead to histone modifications and DNA methylation in metabolic tissues and further contribute to the diabetes risk

Key Concepts

- Metabolic pathways communicate with chromatin and provide information about nutrient availability and energy status via the levels of key metabolites.
- The metabolites acetyl-CoA, α-KG, NAD$^+$, FAD, ATP or SAM act as co-factors and substrates of chromatin modifiers, such as KMTs, KATs, KDACs and TET enzymes.
- The methyl donor substrate SAM connects the processes intermediary metabolism and DNA methylation.
- Chromatin modifiers can act as sensors of the metabolic status of a cell and translate the metabolic information into dynamic post-translational histone modifications that coordinate adaptive transcriptional responses.
- A wide spectrum of secondary metabolites that humans take up from fruits, vegetables, teas, spices, as well as from traditional medicinal herbs, such as genistein, resveratrol, curcumin and polyphenols from green tea, coffee and cocoa, are all able to modulate the activity of chromatin modifiers.
- Gene expression programs that control cell fate decisions are based on epigenetic events that be memorized in the epigenome of metabolic organs.
- Insults to the energy status of the cell are memorized on the level of histone modifications and can be translated into functional outputs via adaptive gene regulation.
- Beneficial effects of calorie restriction on metabolic health are based on NAD$^+$-stimulated sirtuins.

(continued)

Key Concepts (continued)

- The acetyl-CoA/CoA ratio is an important regulator of major cellular decisions.
- The lifestyle of individuals, i.e. primarily their daily diet, creates a metabolic memory.
- Epigenetic programing during embryogenesis, but also in adult life, may explain the missing genetic heritability of the susceptibility for complex diseases, such as T2D.
- Changes in diet or in metabolism, being associated with obesity, can cause an epigenetic drift that may be inherited to the following generations.
- There are new therapeutic strategies, such as a possible reprograming of the epigenome of metabolic organs through personalized diet, that modulate the activity of chromatin modifiers and transcription factors.

Additional Reading

Avgustinova A, Benitah SA (2016) Epigenetic control of adult stem cell function. Nat Rev Mol Cell Biol 17:643–658

Barres R, Zierath JR (2016) The role of diet and exercise in the transgenerational epigenetic landscape of T2DM. Nat Rev Endocrinol 12:441–451

Carlberg C, Ulven SM, Molnár F (2016) Nutrigenomics. Switzerland: Springer Textbook. ISBN: 978-3-319-30413-7

Gut P, Verdin E (2013) The nexus of chromatin regulation and intermediary metabolism. Nature 502:489–498

Kinnaird A, Zhao S, Wellen KE et al (2016) Metabolic control of epigenetics in cancer. Nat Rev Cancer 16:694–707

Sales VM, Ferguson-Smith AC, Patti ME (2017) Epigenetic mechanisms of transmission of metabolic disease across generations. Cell Metab 25:559–571

Glossary

Assay for transposase accessible chromatin using sequencing (ATAC-seq)
Similarly to DNase hyper-sensitivity mapping, this method is used to identify active regulatory sites characterized by lower density of nucleosomes. It uses the Tn5 transposase, which – owing to steric hindrance – can insert sequencing adaptor sequences only into regions free of nucleosomes.

Autistic spectrum disorders a group of neurodevelopmental diseases characterized by deficits in social and communicative interaction and stereotypic behaviors.

Beckwith–Wiedemann syndrome A predominantly maternally transmitted disorder, involving fetal and post-natal overgrowth and a predisposition to embryonic tumors. The Beckwith–Wiedemann syndrome locus includes several imprinted genes, including *IGF2*, *H19* and *KCNQ1*, and loss of imprinting at *IGF2* is seen in ~20% of cases.

Bisulfite sequencing An assay to study 5mC DNA methylation. Native DNA is exposed to bisulfite. Unmethylated cytosines undergo deamination and are converted to uracils, which are read as thymines, whereas methylated cytosines remain unconverted. Sequencing libraries are generated from the converted template and they allow the study of methylation at single-base resolution.

Bivalent domains Chromatin regions that harbor 'active' and 'repressive' histone modifications. Bivalent domains are thought to mark genes that are expressed at low levels only, but that are poised for activation upon a differentiation cue or other signaling events.

Blastocysts Early stage embryos that have undergone the first cell lineage specification, which results in two primary cell types: cells of the inner cell mass and the trophoblasts.

Bromodomain A protein module of ~110 amino acids that mediates interaction with acetylated lysines and is often found in KATs and ATP-dependent chromatin remodeling factors.

© Springer Nature Singapore Pte Ltd. 2018
C. Carlberg, F. Molnár, *Human Epigenomics*, https://doi.org/10.1007/978-981-10-7614-5

Cap analysis of gene expression (CAGE) Capture of the methylated cap at the 5'-end of mRNA, followed by high-throughput sequencing of a small tag (~20 nt) adjacent to the 5'-methylated caps. 5'-methylated caps are formed at the initiation of transcription, although other mechanisms also methylate 5'-ends of RNA.

Cellular reprograming Conversion of a differentiated cell to an embryonic state.

Cell states The transcriptional output of a gene regulatory network, with a variable degree of stability; development is characterized by sequences of cell states that culminate in specific fates.

Chromatin immunoprecipitation followed by sequencing (ChIP-seq) A method for genome-wide mapping of the distribution of histone modifications and chromatin associated proteins that relies on immunoprecipitation with antibodies to modified histones or other chromatin proteins. The enriched DNA is sequenced to create genome-wide profiles.

Chromatin conformation capture (3C) An assay for studying chromosomal 3D structure by proximity ligation. The assay relies on cross-linking chromatin with a fixing agent (usually formaldehyde), digestion of the DNA with a six-base or four-base cutter restriction enzyme and, finally, ligation of the fixed chromatin. In the resulting chimeric DNA template, regions that were close spatially are now closed linearly.

Chromatin interaction analysis with paired-end tag sequencing (ChIA-PET) A high-throughput method based on a combination of ChIP and chromatin proximity ligation assays to predict long-range chromatin interactions that are mediated by either Pol II or transcription factors.

Chromodomain A modular methyl-binding domain of 40–50 amino acids that is commonly found in proteins involved in chromatin remodeling.

Clustered regularly interspaced short palindromic repeats-CRISPR-associated protein 9 (CRISPR-Cas9) Components of a bacterial defense system against viruses are used in this method for the editing mammalian genes.

Constitutive heterochromatin A subtype of heterochromatin that is present at the highly repetitive DNA sequences found at the centromeres and telomeres of chromosomes, where it hinders transposable elements from becoming activated and thereby ensures genome stability and integrity.

CpG island Region of several hundred to approximately 2,000 bp that are frequently found at promoter regions and that exhibit strong enrichment for CpG dinucleotides. CpG islands at promoters are predominantly unmethylated across cell types.

Ectoderm The outermost layer of the three embryonic germ layers that gives rise to the epidermis (for example, the skin, hair and eyes) and the nervous system.

Embryonic stem (ES) cell A type of pluripotent stem cell that is derived from the inner cell mass of the early embryo. Pluripotent cells are capable of generating virtually all cell types of the organism.

Endoderm The innermost layer of the three embryonic germ layers that gives rise to the epithelia of the digestive and respiratory systems, such as liver, pancreas and lungs.

Epigenetic stochasticity Non-deterministic changes of epigenetic marks such as DNA methylation, giving rise to epigenetic variation that underlies cellular plasticity in both normal and pathological states, and that can be localized to specific genomic regions.

Epigenome-wide association studies (EWASs) Studies of the epigenome for nonrandom association of a difference in organization of a genomic regulator, comparing individuals with a phenotype with individuals lacking the phenotype. The epigenome is itself defined as the genome-wide distribution of transcriptional regulators believed to mediate the memory of past cellular events.

Epimutation Heritable stochastic change in the chromatin state at a given position or region. In the context of cytosine methylation, epimutations are defined as heritable stochastic changes in the methylation status of a single cytosine or of a region or cluster of cytosines. Such changes do not necessarily imply changes in gene expression.

Erasers Enzymes that remove histone modifications from chromatin.

Euchromatin Light-staining, decondensed and transcriptionally accessible regions of the genome.

Facultative heterochromatin A subtype of heterochromatin that is formed in the euchromatic environment, where heterochromatin proteins are used to stably repress the activity of certain target genes.

Formaldehyde Assisted Isolation of Regulatory Elements followed by sequencing (FAIRE-seq) This method exploits the solubility of open chromatin in the aqueous phase during phenol-chloroform extraction to generate genome-wide maps of soluble chromatin.

Gene regulatory networks (GRNs) GRNs represent units of interacting proteins that are functionally constrained by defined regulatory relationships. These interactions provide a structure and determine an output in the form of a pattern of gene expression. GRNs are usually represented by nodes (proteins) and edges (their interactions).

Genome The complete haploid DNA sequence of an organism comprising all coding genes and far larger non-coding regions. With the exception of cancer cells, the genome (3,260 Mb) of all of the 400 tissues and cell types that form a human individual is identical and constant over time.

Genome-wide association studies (GWASs) Studies that aim to identify genetic loci associated with an observable trait, disease or condition.

Genomic imprinting An epigenetic phenomenon in which expression of a gene is restricted to a single allele based on parental origin.

Heterochromatin Dark-staining, condensed and gene-poor regions of the genome.

Hi-C A 3C-based method for genome-wide analysis of chromosome conformation. Hi-C involves deep sequencing of chimeric 3C DNA templates and subsequent statistical analysis of the distribution of ligation junctions over two-dimensional contact matrices. Ultra-deep sequencing, or variants of Hi-C that involve enrichment for specific regions of interest, can be used to enhance the assay's resolution.

Implantation An early developmental stage at which the embryo adheres to the endometrium.

Imprinting A chromatin state defined by whether the gene or genetic locus is inherited from the male or the female germ line.

Inner cell mass A group of cells inside a mammalian blastocyst that gives rise to the embryo.

Insulator A chromatin element that acts as a barrier against the influence of positive signals (from enhancers) or negative signals (from silencers and heterochromatin).

Mesoderm The middle layer of the three embryonic germ layers that gives rise to the muscle, cartilage, bone, blood, connective tissue, etc.

Multipotent Pertaining to the ability of a cell to differentiate into multiple but a limited range of cell types (e.g., cells of the embryonic germ layers and adult stem cells are multi-potent).

Odds ratio The mathematical expression of the relation between the presence or absence of a variant (e.g., a particular allele of a SNP) and the presence or absence of a trait (e.g., cancer) in the population. The odds of occurrence of the variant is determined in groups of subjects with and without the trait; the odds ratio then is the ratio of the odds of occurrence of the variant in people with the trait to the odds of occurrence of the variant in people without the trait.

Plasticity The reversibility of epigenetic marks on DNA and proteins.

Pleiotropic Genetic or epigenetic changes that affect multiple seemingly unrelated phenotypic traits.

Pluripotency The ability of a cell to differentiate into all three germ layers and to give rise to all fetal or adult cell types (e.g., cells of the inner cell mass of blastocyst stage embryos are pluripotent).

Poised enhancer An inactive status of an enhancer carrying histone markers, such as H3K4me1, that enables after appropriate stimulation the rapid re-activation of the enhancer.

Post-translational modifications Only after acetylation, methylation, phosphorylation and other covalent modifications most proteins reach their full functional profile. Due to post-translational modifications the proteome is far more complex than the transcriptome and also varies a lot in response to extra- and intra-cellular signals.

Proteome In analogy to the transcriptome, the proteome is the complete set of all expressed proteins in a given tissue of cell type. The proteome depends on the transcriptome, but is not its 1:1 translation, that is transcriptome analyses provide only a very rough description of the resulting proteome.

Quantitative trait loci (QTLs) Loci in the genome at which genetic variation is associated with molecular variation across individuals. For example, individuals with a particular single nucleotide variant have altered expression levels of a gene (eQTL), altered DNA methylation (meQTL; also known as mQTL) or altered chromatin state (chromQTL).

Readers Nuclear proteins that recognize and bind chromatin through histone modification recognition domains.

RNA sequencing (RNA-seq) Isolation of RNA sequences, often with different purification techniques to isolate different fractions of RNA followed by high-throughput sequencing.

Silencers DNA sequences that cause reduced expression of their target gene(s).

SWI/SNF complex A protein complex with ATPase activity that uses the energy of ATP hydrolysis to mobilize nucleosomes and remodel chromatin.

TET family The ten-eleven translocation family of α-KG-dependent dioxygenases that catalyze the oxidation of 5mC to 5hmC and further products. Genes encoding these enzymes are frequently mutated in human cancers.

Topologically associated domains (TADs) Large genomic regions promoting regulatory interactions by forming higher-order chromatin structures separated by boundary regions.

Totipotent Pertaining to the ability of a cell to give rise to differentiated cells of all tissues, including embryonic and extra-embryonic tissues, in an organism (e.g., a zygote is totipotent).

Transcription factor binding motif A degenerate short (6–10 bp) DNA sequence pattern that summarizes the DNA sequence binding preference of a transcription factor. These motifs are usually represented as sequence logos based on position weight matrices.

Transcription start sites (TSSs) Nucleotides in the genome that are the first to be transcribed into a particular RNA.

Transcriptome The complete set of all transcribed RNA molecules of a tissue or cell type. It significantly differs between tissues and depends on extra- and intra-cellular signals.

Transgenerational inheritance Transmission of epigenetic information that is passed on to gametes without alteration of the DNA sequence.

Trithorax group (TrxG) protein Proteins that belong to this family form large complexes and maintain the stable and heritable expression of certain genes throughout development.

Trophoblast The outer layer of the mammalian blastocyst that eventually develops to form part of the placenta.

Waddington landscape A metaphor of development, in which valleys and ridges illustrate the epigenetic landscape that guides a pluripotent cell to a well-defined differentiated state, represented by a ball rolling down the landscape.

Writers Enzymes that add histone modifications to chromatin.

X chromosome inactivation A process in which one of the two X chromosomes is randomly inactivated in female mammalian cells early in development.

Zygote The fertilized egg before cleavage occurs; that is, the one-cell stage embryo.

Printed in the United States
By Bookmasters